FIFTH MATHEMATICAL CONGRESS

COMPTES RENDUS

du

CINQUIÈME CONGRÈS CANADIEN DE MATHÉMATIQUES

Université de Montréal, 1961

Edité par

E. M. ROSENTHALL
Professeur de Mathématiques, McGill University

UNIVERSITY OF TORONTO PRESS
Toronto, Canada
1963

PROCEEDINGS

of the

FIFTH CANADIAN MATHEMATICAL CONGRESS

University of Montreal, 1961

Edited by

E. M. ROSENTHALL

Professor of Mathematics, McGill University

UNIVERSITY OF TORONTO PRESS

Toronto, Canada

1963

Printed in Canada
Reprinted in 2018
ISBN 978-1-4875-7330-0 (paper)

PRÉFACE

LE PRÉSENT OUVRAGE constitue un compte rendu du huitième séminaire biennal et du cinquième congrès de la Société Mathématique du Canada tenus à l'Université de Montréal durant l'été 1961. Le congrès de fondation de notre Société eut lieu à Montréal en 1945 à l'Université McGill. Dans la préface des Actes de ce premier congrès, nous pouvons lire : «Nous souhaitons vivement que cette Société soit à l'origine d'importants développements mathématiques au Canada.»

Le présent rapport démontre amplement que ce voeu a été réalisé bien au-delà des plus optimistes espérances. Dès le début, la Société s'est proposé de couvrir tout le domaine de l'activité mathématique. Une aide importante pour y parvenir fut la collaboration de la Société de Physique Théorique dans l'organisation de sessions conjointes. Les détails du programme de cette session 1961 révèlent l'étendue du succès obtenu. Comme aux sessions précédentes, des conférences au niveau de la recherche sont données par d'éminents professeurs étrangers. Il y a aussi des séries de cours supérieurs donnés généralement par des canadiens. Le programme offre également plusieurs points d'intérêt pour les professeurs de l'enseignement secondaire. On y trouve en particulier un symposium sur la formation dans les universités des futurs professeurs d'enseignement secondaire ainsi que des discussions sur le curriculum en mathématiques et en physique au niveau secondaire. Ce programme, à l'instar des programmes des sessions antérieures, reflète les multiples activités de la Société et témoigne du niveau de l'activité mathématique générale au Canada. Parmi les activités de la Société mentionnons : l'octroi de bourses aux étudiants candidats à la maîtrise ou au doctorat ainsi qu'aux étudiants post-doctoraux, l'Institut de recherches d'été, les sessions de cours d'été à l'intention des professeurs de mathématiques du niveau secondaire, les concours dans la plupart des provinces du Canada avec un programme de prix et de bourses à l'intention des étudiants finissants des écoles secondaires.

Durant ces sessions, des conférences et des cours sont donnés à tous les niveaux dans les deux langues officielles. Dès le début, la Société s'est proposé de tirer parti du caractère bi-culturel de notre pays. Les représentants des deux cultures française et anglo-saxonne ont coopéré avec intérêt et satisfaction à cette fin.

L'Exécutif et les membres de la Société ainsi que tous les invités et visiteurs profitent de l'occasion pour exprimer leur gratitude envers les autorités de l'Université de Montréal pour l'invitation à tenir la session de 1961 sur leur campus. Cette expression de gratitude est également transmise au Comité local d'organisation dont le travail soigneux et efficace contribua pour une large part au succès de cette session.

R.L.J.

PREFACE

THESE PROCEEDINGS report the fifth seminar and general assembly of the Canadian Mathematical Congress, with the University of Montreal as host. The first such meeting was in 1945, and it too was in Montreal with McGill University as host. In the preface to the Proceedings of the first Congress assembly there is the following statement: "It is the hope and belief of many that this Congress will be the beginning of important mathematical developments in Canada."

In the present report, which covers the return meeting to Montreal in 1961, there is ample evidence that this hope has been realized far beyond the most optimistic expectations. From the first it has been the aim of the Congress to encompass the full range of mathematical activity. An important aid in accomplishing this has been the co-operation of the Theoretical Physics Society in arranging joint sessions. The extent to which the over-all aim has been reached is revealed in the details of the programme of the session under review. In both Mathematics and Theoretical Physics there are, as in previous sessions, top level research lectures given by prominent scientists from abroad; there are series of lectures at the post-graduate level, usually given by Canadians; always there are features of interest to those teaching in secondary schools which, in the session under review, take the form of panel discussions on the training at the University given to high school teachers, and discussions on curriculum content in mathematics and physics at the secondary school level. This programme, like that of previous sessions, reflects the many interim activities of the Congress, which in turn reflects the level of activity in mathematics throughout the country.These activities include the granting of fellowships and travelling allowances for graduate students, and for post-doctoral study; the summer research institute; summer sessions for the advancement of high school teachers in subject matter; a nation-wide programme of prize and scholarship competitions at the matriculation level.

At all levels there are lectures in both official languages. From the first it has been the aim of the Congress to make the most of our good fortune as a nation in having our social structure based on two cultures, French and Anglo-Saxon. The representatives of both cultures have co-operated with interest and satisfaction towards this end.

The Executive and members of the Congress, and all invited guests and visitors, take this opportunity to express gratitude to the Officers of the University of Montreal for the invitation to hold the 1961 session on their campus. This expression of appreciation and gratitude is also extended to the Committee on Arrangements for the careful planning which added so much to the success of the session.

R.L.J.

TABLE DES MATIÈRES ||| CONTENTS

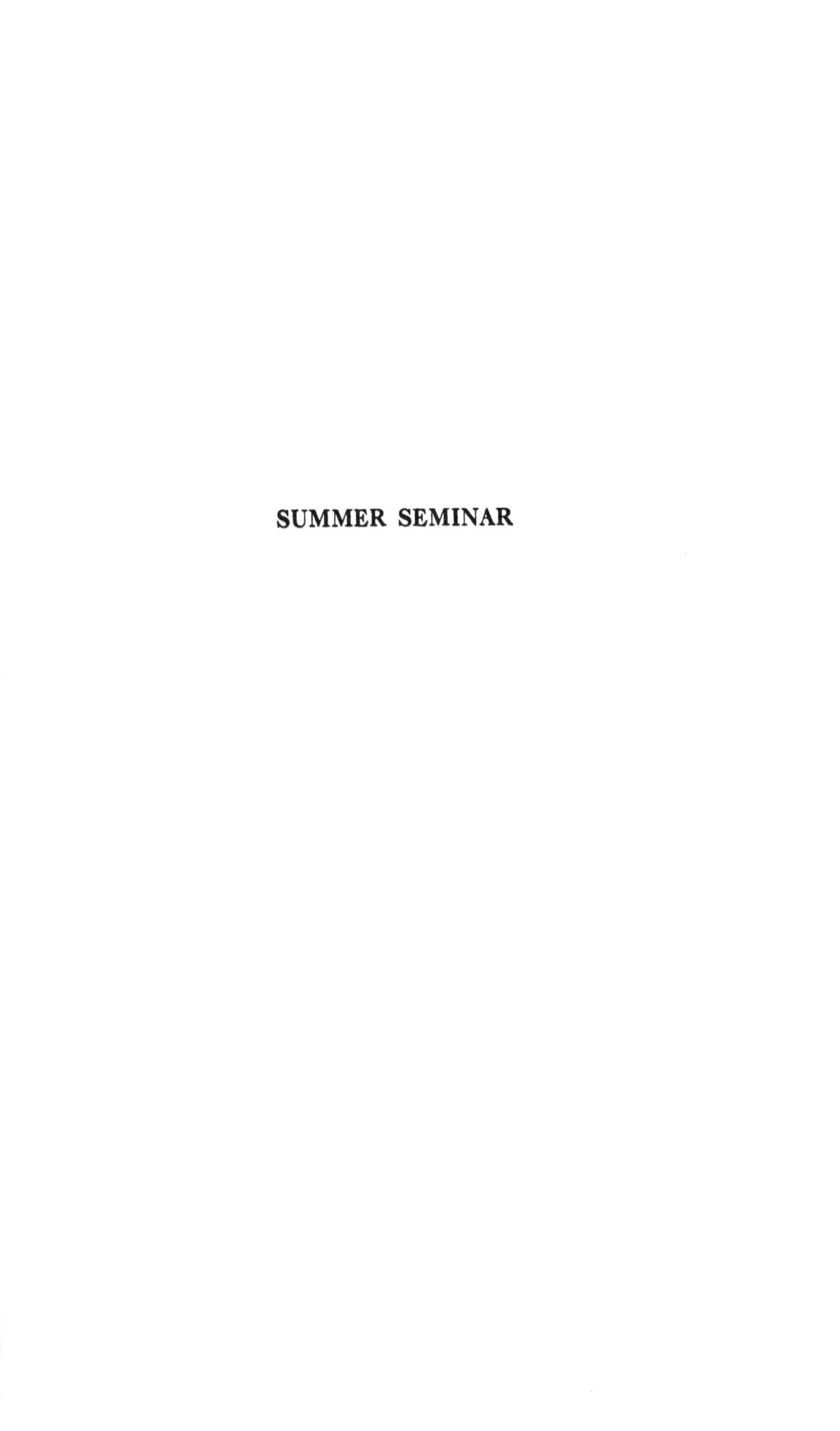

SUMMER SEMINAR

SUMMER SEMINAR

THE SEMINAR OF 1961 was held at the University of Montreal from August 13 to September 1, jointly with the Theoretical Physics Division of the Canadian Association of Physicists. In accordance with the plan followed at earlier seminars, there were series of lectures at two levels, known as the Research Lectures and the Instructional Lectures, in order that the needs of all might be met. The programme was as follows:

RESEARCH LECTURES

C. Ehresmann, *Institut Henri Poincaré, Paris*	Catégories différentiables et géométrie différentielle
A. Erdélyi, *California Institute of Technology*	Asymptotic solutions of differential equations
L. Henkin, *University of California*	The theory of cylindrical algebras
I. N. Sneddon, *The University, Glasgow*	Mixed boundary value problems in potential theory

INSTRUCTIONAL LECTURES

I. Halperin, *Queen's University*	Continuous geometry
T. E. Hull, *University of British Columbia*	Numerical analysis
J. Maranda, *University of Montreal*	Algèbre homologique
H. F. Trotter, *Queen's University*	Probability theory

ENTERTAINMENT

The city of Montreal and the surrounding district proved to be an interesting place for visitors from other centres. Among the social events were a reception for members and their wives given by the University of Montreal, a tea for the wives at McGill University Faculty Club, and a visit to the Morgan Arboretum at Ste. Anne de Bellevue. A picnic at the Arboretum was arranged through the kindness of the staff of Macdonald College at Ste. Anne de Bellevue.

Members of the seminar and their wives were also entertained at a dinner given by the Collège Militaire Royal de Saint Jean at

Saint Jean, Quebec. This was followed by an inspection of the new buildings at the Collège.

Coffee was served each day from Monday through Friday at the University of Montreal, providing an additional opportunity for members of the seminar to become better acquainted and to exchange ideas on their current study and research.

LIST OF THOSE ATTENDING THE SEMINAR

Abramowich, J.	Assumption University, Windsor
Adshead, J. G.	Dalhousie University, Halifax
Anctil, J. A.	Collège Militaire Royal, St. Jean, Québec
Anctil, Mrs. J. A.	
Antliff, W. B.	Royal Military College, Kingston
Artiaga, Lucio	University of Saskatchewan, Saskatoon
Ayoub, R. G.	Pennsylvania State University
Ayoub, Mrs. Christine	Pennsylvania State University
Baragar, Diana E.	McMaster University, Hamilton
Beaudet, Gilles	University of Montreal
Benjamin, I.	Loyola College, Montreal
Bonyun, D. A.	McGill University, Montreal
Bonyun, Mrs. D. A.	
Bouwer, I. Z.	Queen's University, Kingston
Brender, Allan	University of California, Berkeley
Brossard, Roland	University of Montreal
Brossard, Mrs. Roland	
Buchal, R. N.	Argonne National Laboratory, Illinois
Bures, D. J.	Queen's University, Kingston
Bush, G. C.	Queen's University, Kingston
Bush, Mrs. G. C.	
Carragher, Paul	Memorial University of Newfoundland
Clement, Bernard	University of Montreal
Cléroux, Robert	University of Montreal
Cole, R. H.	University of Western Ontario, London
Cole, Mrs. R. H.	
Courteau, Bernard	University of Montreal
Crawford, W. S. H.	Mount Allison University, Sackville
Crosby, D. R.	University of Alberta, Edmonton
Daigneault, A.	University of Montreal
Daigneault, Mrs. A.	
Dubuc, Serge	University of Montreal
Ehresmann, Charles	Institut Henri Poincaré, Paris
Eliopoulos, H. A.	Assumption University, Windsor
Erdélyi, A.	California Institute of Technology, Pasadena
Evans, Arwel	McGill University, Montreal
Evans, Mrs. Arwel	
Fieldhouse, David	Princeton University, New Jersey

Fortin, Jacques	Laval University, Quebec
Fox, Charles	McGill University, Montreal
Fox, Mrs. Charles	
Fox, G. E. N.	University of Montreal
Gagnon, F. P.	University of Montreal
Gagnon, Mrs. F. P.	
Gandhi, S. R.	University of Western Ontario, London
Gardner, R. B.	University of California, Berkeley
Gauthier, Abel	University of Montreal
Gauthier, Mrs. Abel	
Grätzer, George	Hungarian Academy, Budapest
Guttman, Irwin	McGill University, Montreal
Halperin, I.	Queen's University, Kingston
Hayes, J. C.	Collège Militaire Royal, St. Jean, Quebec
Henkin, Leon	University of California, Berkeley
Henkin, Mrs. Leon	
Hershorn, Michael	McGill University, Montreal
Holbrook, J. A.	Queen's University, Kingston
Hull, Thos. E.	University of British Columbia, Vancouver
Jeffery, R. L.	Acadia University, Wolfville
Kaller, Cecil Louis	University of Saskatchewan, Saskatoon
Kassimatis, C.	University of North Carolina, Raleigh
Katz, I. N.	A.V.C.O., Massachusetts
Katz, Mrs. I. N.	
Kaufman, H.	McGill University, Montreal
Kaufman, Mrs. H.	
Kayler, Hélène	University of Montreal
Keeping, E. S.	University of Alberta, Edmonton
Kemp, R. R. D.	Queen's University, Kingston
Kimura, Naoki	University of Saskatchewan, Saskatoon
Klamkin, M. S.	A.V.C.O., Massachusetts
Kleisli, H.	University of Montreal
Klemola, M. T.	Assumption University, Windsor
Klemola, Mrs. M. T.	
Knight, William	University of New Brunswick, Fredericton
Knight, Mrs. William	
Kotzé, W. J.	McGill University, Montreal
Kobayashi, S.	University of British Columbia, Vancouver
L'Abbé, Maurice	University of Montreal
L'Abbé, Mrs. Maurice	
Lachapelle, Benoit	University of Montreal
Lachapelle, Mrs. Benoit	
Lambek, Joachim	McGill University, Montreal
Lambek, Mrs. Joachim	
Lapointe, Claude	University of Montreal
Leah, Rev. Philip J.	St. Paul's College, Winnipeg
LeBlanc, Léon	University of Montreal
LeBlanc, Mrs. Léon	
Leduc, Pierre-Yves	University of Montreal
Legoupil, Jean	University of Montreal

Lorch, Lee	University of Alberta, Edmonton
MacPhee, Rev. H. J., s.j.	Loyola College, Montreal
Mandl, Dr. Paul	National Research Council
Maranda, Jean	University of Montreal
Maranda, Mrs. Jean	
Melamed, Samuel	McGill University, Montreal
Melamed, Mrs. Samuel	
Melzak, Z. A.	University of British Columbia, Vancouver
Ménard, Jean	University of Montreal
Messerli, R.	University of Montreal
Miller, D. R.	University of Toronto
Mohanty, S. G.	University of Alberta, Edmonton
Munier, François	Ecole Polytechnique, Montreal
Normand, G.	Collège Militaire Royal, St. Jean, Quebec
Normand, Mrs. G.	
O'Connor, Rev. R. E., s.j.	Loyola College, Montreal
Peck, J. E. L.	University of Alberta, Calgary
Pillow, A. F.	University of Toronto
Pillow, Mrs. A. F.	
Pilon, R. G.	Collège Militaire Royal, St. Jean, Quebec
Pilon, Mrs. R. G.	
Pouliot, Adrien	Laval University, Quebec
Prillo, A. A. J.	Loyola College, Montreal
Ritcey, L. F. S.	Canadian Mathematical Congress, Montreal
Ritcey, Mrs. L. F. S.	
Robert, Pierre	University of Montreal
Robert, Mrs. Pierre	
Robillard, Paul	University of Montreal
Rogers, C. A.	University of London, England
Rogers, Mrs. C. A.	
Ronveaux, André	University of Montreal
Rothberger, Fritz	Laval University, Quebec
Schremp, Dr. E. J.	U.S. Naval Research Laboratory, Washington
Schremp, Mrs. E. J.	
Schwerdtfeger, Hans	McGill University, Montreal
Schwerdtfeger, Mrs. Hans	
Shklov, Nathan	University of Saskatchewan, Saskatoon
Shklov, Mrs. Nathan	
Shtern, I. H.	McGill University, Montreal
Shtern, Mrs. I. H.	
Smith, P. D. P.	Memorial University of Newfoundland
Sneddon, I. N.	The University, Glasgow
Sneddon, Mrs. I. N.	
Stallard, S.	Collège Militaire Royal, St. Jean, Quebec
Sussman, David	McGill University, Montreal
Sussman, Mrs. David	
Svilokos, Nikola	Long Island University, Brooklyn
Thierrin, G.	University of Montreal
Thierrin, Mrs. G.	
Thomas, G. H. M.	University of Saskatchewan, Saskatoon

Tomiuk, Daniel	University of Ottawa
Trotter, Hale	Queen's University, Kingston
Turgeon, Rev. Jean, s.j.	Loyola College, Montreal
Van Der Merwe, D. F.	McGill University, Montreal
Venne, Marc	University of Montreal
Wasan, M. T.	Queen's University, Kingston
Watson, R. E.	Southern Illinois University
Watson, Mrs. R. E.	
Williams, W. L. G.	McGill University, Montreal (retired)
Williams, Mrs. W. L. G.	
Wonenburger, Maria J.	Queen's University, Kingston
Woonton, G. A.	McGill University, Montreal
Wyman, Max	University of Alberta, Edmonton
Wyman, Mrs. Max	
Zassenhaus, Hans	Notre Dame University, South Bend, Indiana

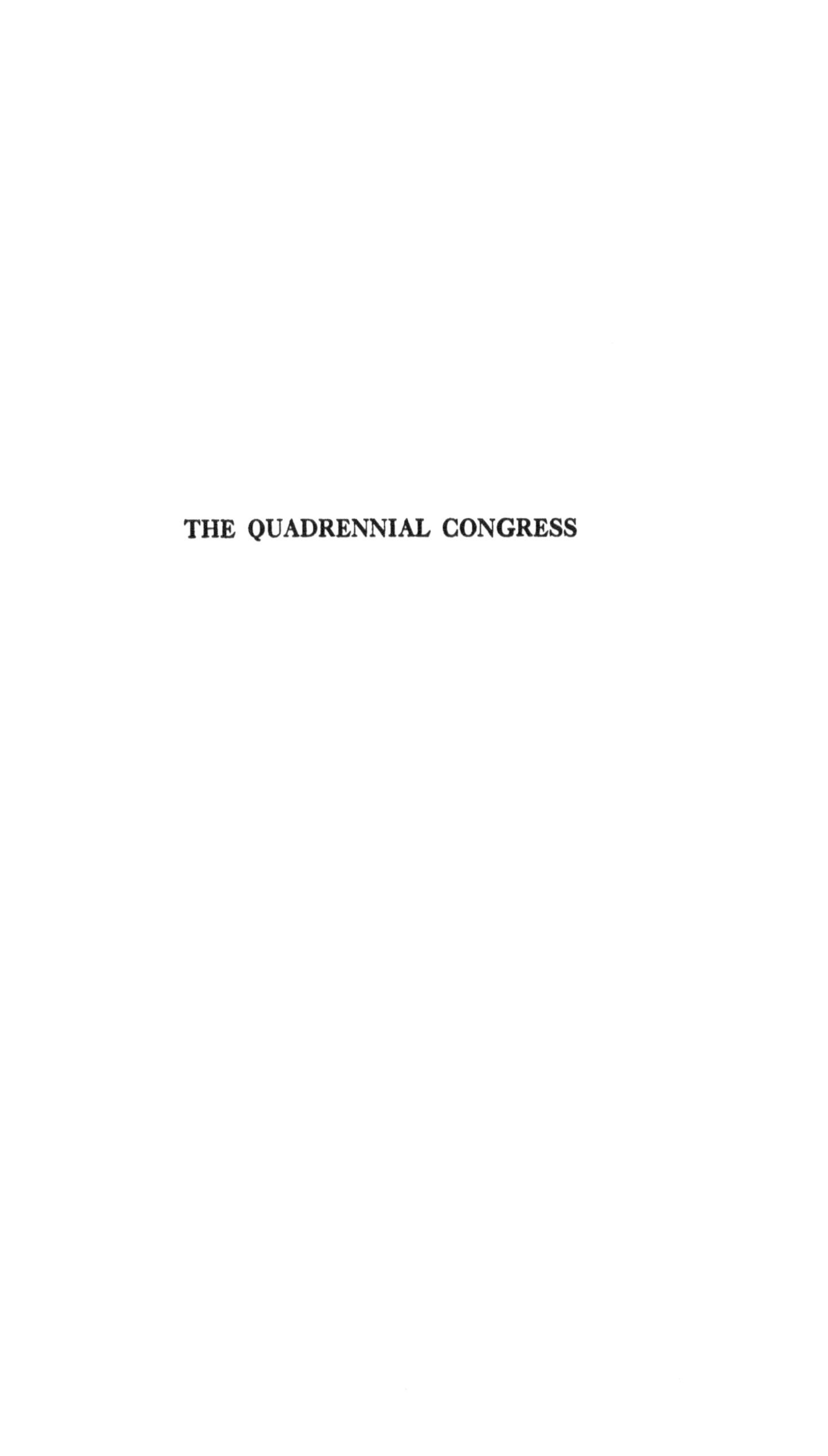

THE QUADRENNIAL CONGRESS

THE QUADRENNIAL CONGRESS OF 1961

THE CONGRESS OF 1961 was held at the University of Montreal from September 3 to September 9, in joint session with the Theoretical Physics Division of the Canadian Association of Physicists. The Congress programme was as follows:

SUNDAY, *September 3*
- 2:00 P.M. Registration.
- 8:00 P.M. Informal Reception.

MONDAY, *September 4*
- 9:30 A.M. Lecture, Henkin.
- 11:00 A.M. Lecture, Chew.
- 2:00 P.M. Business meeting of the Canadian Mathematical Congress.
- 5:00 P.M. Address of the retiring president of the Canadian Mathematical Congress, R. L. Jeffery: "Derivatives and integrals with respect to a base function."

TUESDAY, *September 5*
- 9:30 A.M. Lecture, Erdélyi.
- 11:00 A.M. Lecture, Ehresmann.
- 2:30 P.M. Lecture, de Groot.
- 4:00 P.M. Lecture, Sneddon.
- 8:00 P.M. Symposium on the teaching of applied mathematics in Canada.

WEDNESDAY, *September 6*
- Picnic at Mount Saint-Hilaire.

THURSDAY, *September 7*
- 9:30 A.M. Research Policy Seminar.
- 2:00 P.M. Lecture, Chew.
- 6:30 P.M. Cocktails and dinner offered by the City of Montreal, at the Restaurant Hélène de Champlain, Ile Ste-Hélène.

FRIDAY, *September 8*
- 9:00 A.M. Contributed papers.
- 2:30 P.M. Symposium on the training in mathematics given at the university to high school teachers.
- 8:00 P.M. Discussion on high school mathematics (in French).

SATURDAY, *September 9*
- 9:00 A.M. Lecture, de Groot.
- 10:15 A.M. Discussion on high school mathematics (in English).

INVITED LECTURES

G. F. Chew, *University of California, Berkeley*	The elementary particle concept: A principle of maximum strength for strong interactions

S. R. de Groot, *University of Leiden, Holland*	The causality principle and the fluctuation dissipation theorem; Rayonnements de noyaux orientés
C. Ehresmann, *Institut Henri Poincaré*	Structures feuilletées
A. Erdélyi, *California Institute of Technology*	An extension of the concept of real number
Leon Henkin, *University of California, Berkeley*	Mathematics and logic
I. N. Sneddon, *The University, Glasgow*	The application of mathematics to biology and medicine

CONTRIBUTED PAPERS

Nicholas Chako, *Queens College* and *Institute of Space Studies, N. Y.*	On conservative dynamical systems and the equation of wave mechanics
G. D. Findlay and and J. Lambek, *McGill University, Montreal*	Multiplicative ideal theory of rings of quotients
K. Hoechsmann, *University of Notre Dame, Indiana*	*p*-Algebras and derivations
P. Mandl, *National Research Council, Ottawa*	Flow of heat from a wedge into a surrounding medium
John Mariani, *Fairleigh Dickinson University*	Homomorphism between groups of contact transformations and corresponding groups of point transformations
T. V. Narayana, *University of Alberta, Edmonton*	A note on weighted compositions
P. Savic and J. W. Gosnell, *National Research Council, Ottawa*	The dynamics of the expanding vapour bubble in a boiling liquid
Anne B. Underhill, *Dominion Astrophysical Observatory, Victoria*	Some methods for computing model stellar atmospheres
V. H. Weston, *University of Michigan*	Atmospheric waves produced by a large explosion in the atmosphere

SYMPOSIUM ON THE TEACHING OF APPLIED MATHEMATICS IN CANADA

P. R. Wallace (presiding), *McGill University*
F. H. Northover, *Carleton University*
A. F. Pillow, *University of Toronto*
I. N. Sneddon, *University of Glasgow*

RESEARCH POLICY SEMINAR

G. de B. Robinson (presiding), *University of Toronto*

SYMPOSIUM ON THE TRAINING IN MATHEMATICS GIVEN AT THE UNIVERSITY TO HIGH SCHOOL TEACHERS

I. Halperin (presiding), *Queen's University*
D. Betts, *University of Alberta*
A. W. Bishop, *Department of Education, Ontario*
R. Stanton, *University of Waterloo*

DISCUSSION ON HIGH SCHOOL MATHEMATICS (IN FRENCH)

Maurice L'Abbé (presiding), *University of Montreal*
I. Halperin, *Queen's University*
L. Henkin, *University of California, Berkeley*
P. Obreanu, *Queen's University*

DISCUSSION ON HIGH SCHOOL MATHEMATICS (IN ENGLISH)

L. F. S. Ritcey (presiding), *Canadian Mathematical Congress*
A. J. Coleman, *Queen's University*
L. Henkin, *University of California, Berkeley*
R. L. Rosenberg, *University of New Brunswick*

ENTERTAINMENT

An informal reception was held at the University of Montreal on Sunday evening, September 3.

On Wednesday, all members of the Congress and their families attended a picnic at Mount St. Hilaire. This property, which belongs to McGill University, is situated twenty miles east of Montreal.

On Thursday evening a dinner was offered by the City of Montreal on St. Helene's Island in the St. Lawrence River. The speakers at this dinner were Professor M. L'Abbé (chairman), Professor R. L. Jeffery, and Professor I. N. Sneddon.

BUSINESS MEETING

A meeting of the Congress was held on Monday in which reports were given and other business was transacted.

A constitution, copies of which had been sent to all members, was adopted by the Congress. According to this constitution, officers will be elected every second year for a two-year period. In future years the election will be held by mail ballot, but in 1961 elections were held at the Congress meeting. The list of elected and permanent officers is as follows:

Past President: R. L. Jeffery, *Acadia University*
President: R. D. James, *University of British Columbia*
Vice-Presidents: A. Gauthier, *University of Montreal*; N. S. Mendelsohn, *University of Manitoba*; R. L. Rosenberg, *University of New Brunswick*
Executive Director: L. F. S. Ritcey
Treasurer: W. L. G. Williams (retired), *McGill University*
English Secretary: R. E. O'Connor, s.j., *Loyola College*
French Secretary: J. M. A. Maranda, *University of Montreal*

Council Members

From the West: R. Blum, *University of Saskatchewan*; D. Derry, *University of British Columbia*; A. P. Guinand, *University of Saskatchewan*; L. Moser, *University of Alberta*; B. N. Moyls, *University of British Columbia*; B. Noonan, *University of Manitoba*; J. E. L. Peck, *University of Alberta at Calgary*; M. Wyman, *University of Alberta*

From Ontario and Quebec: B. Banaschewski, *McMaster University*; R. H. Cole, *University of Western Ontario*; A. J. Coleman, *Queen's University*; H. S. M. Coxeter, *University of Toronto*;

Rev. D. T. Faught, *Assumption University*; K. D. Fryer, *University of Waterloo*; J. Lambek, *McGill University*; N. D. Lane, *McMaster University*; V. Linis, *University of Ottawa*; F. Rothberger, *Laval University*; E. M. Rosenthall, *McGill University*; H. Schwerdtfeger, *McGill University*

From the Atlantic Provinces: W. J. Blundon, *Memorial University of Newfoundland*; W. S. H. Crawford, *Mount Allison University*; D. O. Snow, *Acadia University*; A. J. Tingley, *Dalhousie University*

Editor-in-Chief, Canadian Journal of Mathematics: W. T. Tutte, *University of Toronto*

Managing Editor, Canadian Journal of Mathematics: G. de B. Robinson, *University of Toronto*

Editor-in-Chief, Canadian Mathematical Bulletin: W. Moser, *University of Manitoba*

Managing Editor, Canadian Mathematical Bulletin; P. Scherk, *University of Toronto*

Director, Summer Research Institute: R. L. Jeffery, *Acadia University*

LIST OF THOSE ATTENDING THE CONGRESS

Abramowich, J.	Assumption University, Windsor
Adshead, J. G.	Dalhousie University, Halifax
Anctil, J. A.	Collège Militaire Royal, St. Jean, Quebec
Anctil, Mrs. J. A.	
Antliff, W. B.	Royal Military College, Kingston
Artiaga, Lucio	University of Saskatchewan, Saskatoon
Ayoub, R. G.	Pennsylvania State University
Ayoub, Mrs. Christine	Pennsylvania State University
Baragar, Diana E.	McMaster University, Hamilton
Beaudet, Gilles	University of Montreal
Benjamin, I.	Loyola College, Montreal
Blundon, W. J.	Memorial University of Newfoundland
Bonyun, D. A.	McGill University, Montreal
Bonyun, Mrs. D. A.	
Bouwer, I. Z.	Queen's University, Kingston
Brender, Allan	University of California, Berkeley
Brossard, Roland	University of Montreal
Brossard, Mrs. Roland	
Buchal, Robert N.	Argonne National Laboratory, I linois
Bures, D. J.	Queen's University, Kingston

Bush, G. C.	Queen's University, Kingston
Bush, Mrs. G. C.	
Carragher, Paul	Memorial University of Newfoundland
Clement, Bernard	University of Montreal
Cléroux, Robert	University of Montreal
Cole, R. H.	University of Western Ontario, London
Cole, Mrs. R. H.	
Coleman, A. J.	Queen's University, Kingston
Courteau, Bernard	University of Montreal
Coxeter, H. S. M.	University of Toronto
Crawford, W. S. H.	Mount Allison University, Sackville
Crosby, D. R.	University of Alberta, Edmonton
Csorgo, M.	McGill University, Montreal
Daigneault, A.	University of Montreal
Daigneault, Mrs. A.	
De Laet, Christian	KCS (Quebec) Limited, Montreal
Derry, D.	University of British Columbia, Vancouver
Derry, Mrs. D.	
Divinsky, Nathan	University of British Columbia, Vancouver
Dubuc, Serge	University of Montreal
Edwards, L. P.	University of New Brunswick, Fredericton
Ehresmann, Charles	Institut Henri Poincaré, Paris
Eliopoulos, H. A.	Assumption University, Windsor
Erdélyi, A.	California Institute of Technology, Pasadena
Evans, Arwel	McGill University, Montreal
Evans, Mrs. Arwel	
Faught, Rev. Donald T.	Assumption University of Windsor
Fieldhouse, David	Princeton University, New Jersey
Fortin, Jacques	Laval University, Quebec
Fox, Charles	McGill University, Montreal
Fox, Mrs. Charles	
Fox, G. E. N.	University of Montreal
Fryer, K. D.	University of Waterloo, Ontario
Gagnon, F. P.	University of Montreal
Gagnon, Mrs. F. P.	
Gandhi, S. R.	University of Western Ontario, London
Gardner, R. B.	University of California, Berkeley
Gauthier, Abel	University of Montreal
Gauthier, Mrs. Abel	
Grätzer, George	Hungarian Academy, Budapest
Guttman, Irwin	McGill University, Montreal
Haley, K. D. C.	Acadia University, Wolfville
Halperin, I.	Queen's University, Kingston
Hayes, J. C.	Collège Militaire Royal, St. Jean, Quebec
Henkin, Leon	University of California, Berkeley
Henkin, Mrs. Leon	
Hershorn, Michael	McGill University, Montreal
Hoechsmann, K.	University of Notre Dame, Indiana
Holbrook, J. A.	Queen's University, Kingston
Hull, Thos. E.	University of British Columbia, Vancouver

James, R. D.	University of British Columbia, Vancouver
Jeffery, R. L.	Acadia University, Wolfville
Kaller, C. L.	University of Saskatchewan, Saskatoon
Kassimatis, C.	University of North Carolina, Raleigh
Katz, I. N.	A.V.C.O., Massachusetts
Katz, Mrs. I. N.	
Kaufman, H.	McGill University, Montreal
Kaufman, Mrs. H.	
Kayler, Hélène	University of Montreal
Keeping, A. J.	University of Alberta, Edmonton
Keeping, E. S.	University of Alberta, Edmonton
Kemp, R. R. D.	Queen's University, Kingston
Kimura, Naoki	University of Saskatchewan, Saskatoon
Klamkin, M. S.	A.V.C.O., Massachusetts
Kleisli, H.	University of Montreal
Klemola, M. T.	Assumption University, Windsor
Klemola, Mrs. M. T.	
Knight, William	University of New Brunswick, Fredericton
Knight, Mrs. William	
Kotzé, W. J.	McGill University, Montreal
Kobayashi, S.	University of British Columbia, Vancouver
L'Abbé, Maurice	University of Montreal
L'Abbé, Mrs. Maurice	
Lachapelle, Benoit	University of Montreal
Lachapelle, Mrs. Benoit	
Lambek, Joachim	McGill University, Montreal
Lambek, Mrs. Joachim	
Lapointe, Claude	University of Montreal
Laufer, P. J.	Collège Militaire Royal, St. Jean, Quebec
Laufer, Mrs. P. J.	
Leah, Rev. Philip J.	St. Paul's College, Winnipeg
LeBlanc, Léon	University of Montreal
LeBlanc, Mrs. Léon	
Leduc, Pierre-Yves	University of Montreal
Legoupil, Jean	University of Montreal
Linis, V.	University of Ottawa
Lorch, Lee	University of Alberta, Edmonton
MacPhee, Rev. H. J., s.j.	Loyola College, Montreal
Magee, G. R.	University of Western Ontario, London
Mandl, Dr. Paul	National Research Council
Maranda, Jean	University of Montreal
Maranda, Mrs. Jean	
Mariani, John	Fairleigh Dickinson University
Melamed, Samuel	McGill University, Montreal
Melamed, Mrs. Samuel	
Melzak, Z. A.	University of British Columbia, Vancouver
Ménard, Jean	University of Montreal
Mendelsohn, N. S.	University of Manitoba, Winnipeg
Messerli, R.	University of Montreal
Miller, C. E.	University of Saskatchewan, Saskatoon

Miller, D. R.	University of Toronto
Mohanty, S. G.	University of Alberta, Edmonton
Moser, W. O.	University of Manitoba, Winnipeg
Munier, François	Ecole Polytechnique, Montreal
Narayana, T. V.	University of Alberta, Edmonton
Normand, G.	Collège Militaire, Royal St. Jean, Quebec
Normand, Mrs. G.	
Obreanu, P.	Queen's University, Kingston
O'Connor, Rev. R. E., s.j.	Loyola College, Montreal
Peck, J. E. L.	University of Alberta, Edmonton
Pillow, A. F.	University of Toronto
Pillow, Mrs. A. F.	
Pilon, R. G.	Collège Militaire Royal, St. Jean, Quebec
Pilon, Mrs. R. G.	
Pouliot, Adrien	Laval University, Quebec
Pounder, I. R.	York University, Toronto
Prillo, A. A. J.	Loyola College, Montreal
Raboy, D.	Tutorial High School, Montreal
Ritcey, L. F. S.	Canadian Mathematical Congress, Montreal
Ritcey, Mrs. L. F. S.	
Robert, Pierre	University of Montreal
Robert, Mrs. Pierre	
Robillard, Paul	University of Montreal
Robinson, G. de B.	University of Toronto
Robinson, Mrs. G. de B.	
Rogers, C. A.	University of London, England
Rogers, Mrs. C. A.	
Ronveaux, André	University of Montreal
Rooney, P. J.	University of Toronto
Rosenberg, R. L.	University of New Brunswick, Fredericton
Rosenthall, E. M.	McGill University, Montreal
Rosenthall, Mrs. E. M.	
Rothberger, Fritz	Laval University, Quebec
Russell, Dennis	York University, Toronto
Russell, Mrs. Dennis	
Scherk, Peter	University of Toronto
Schirmer, Dr. H. H.	University of New Brunswick, Fredericton
Schremp, Dr. E. J.	U.S. Naval Research Laboratory, Washington
Schremp, Mrs. E. J.	
Schwerdtfeger, Hans	McGill University, Montreal
Schwerdtfeger, Mrs. Hans	
Shklov, Nathan	University of Saskatchewan, Saskatoon
Shklov, Mrs. Nathan	
Shtern, I. H.	McGill University, Montreal
Shtern, Mrs. I. H.	
Smith, P. D. P.	Memorial University of Newfoundland
Sneddon, I. N.	The University, Glasgow
Sneddon, Mrs. I. N.	
Snow, D. O.	Acadia University, Wolfville
Stallard, S.	Collège Militaire Royal, St. Jean, Quebec

Stanton, R. G.	University of Waterloo, Ontario
St. Pierre, Jacques	University of Montreal
Sumner, D. B.	McMaster University, Hamilton
Sussman, David	McGill University, Montreal
Sussman, Mrs. David	
Svilokos, Nikola	Long Island University, Brooklyn
Thierrin, G.	University of Montreal
Thierrin, Mrs. G.	
Thomas, G. H. M.	University of Saskatchewan, Saskatoon
Tingley, Arnold J.	Dalhousie University, Halifax
Tomiuk, Daniel	University of Ottawa
Trotter, Hale	Queen's University, Kingston
Turgeon, Rev. Jean, s.j.	Loyola College, Montreal
Tutte, W. T.	University of Toronto
Underhill, Anne B.	Dominion Astrophysical Observatory, Victoria
Van Der Merwe, D. F.	McGill University, Montreal
Venne, Marc	University of Montreal
Wallace, P. R.	McGill University, Montreal
Wallace, Mrs. P. R.	
Wasan, M. T.	Queen's University, Kingston
Watson, R. E.	Southern Illinois University
Watson, Mrs. R. E.	
Williams, W. L. G.	McGill University, Montreal (retired)
Williams, Mrs. W. L. G.	
Wittenberg, A.	Laval University, Quebec
Wonenburger, Maria J.	Queen's University, Kingston
Woonton, G. A.	McGill University, Montreal
Wyman, Max	University of Alberta, Edmonton
Wyman, Mrs. Max	
Zassenhaus, Hans	Notre Dame University, South Bend, Indiana

ACKNOWLEDGMENT

As Executive Director of the Canadian Mathematical Congress, I take pleasure in here recording, on behalf of the Congress, the names of the public bodies and companies which contributed to our general fund during the year 1961, and thus gave their support to our various activities.

The present list includes not only the National Research Council and five provinces but also almost all of the Canadian life assurance companies and chartered banks, and a large number of industrial and engineering companies.

Sustaining Members

Province of Nova Scotia
Province of Ontario
Province of Quebec

Bank of Montreal
British Columbia Electric Company Limited
Canada Life Assurance Company
Canadian Industries, Limited
Consolidated Mining and Smelting Company of Canada, Limited
Great-West Life Assurance Company
Hydro-Electric Power Commission of Ontario
Imperial Oil Limited
International Nickel Company of Canada, Limited
London Life Insurance Company
Manufacturers Life Insurance Company
Northern Electric Company, Limited
Royal Bank of Canada
Shawinigan Water and Power Company and associated companies
Steel Company of Canada
Sun Life Assurance Company of Canada
Winnipeg Actuaries Club (funds provided by the life insurance companies wtih Head Offices in Winnipeg)

Other Contributors

National Research Council
Province of New Brunswick
Province of Saskatchewan
Abitibi Power and Paper Company, Limited
Algoma Steel Corporation, Limited
Alliance Mutual Life Insurance Company
Allied Chemical Canada, Limited
Allsopp and Company, Edmonton
Aluminum Company of Canada Limited
A. E. Ames and Company Limited
A. M. F. Atomics, Limited, Port Hope
Anthes-Imperial Company Limited, St. Catharines
Aronovitch and Leipsic Limited, Winnipeg
J. H. Ashdown Hardware Company Limited
L'Assurance Vie Desjardins
Babcock-Wilcox and Goldie-McCulloch Limited, Galt
Bach-Simpson Limited, London
Bailey-Selburn Oil and Gas Ltd., Calgary
Bank of Nova Scotia
Banque Canadienne Nationale
B. P. Baragar and Company Ltd., Edmonton
S. Belzberg Enterprises, Edmonton
Beneficial Finance Company of Canada
Henry Birks and Sons (Montreal) Limited
British American Oil Company Limited
British Columbia Forest Products Limited, Vancouver
Brody's Limited, Edmonton
Bruck Mills Limited, Montreal
Buckwold's Limited, Saskatoon
Honourable C. J. Burchell, Halifax
Burns and Company, Limited
Calgary Power, Limited
California Standard Company, Calgary
Canada Cement Company Limited
Canada Iron Foundries Limited
Canada Packers Foundation
Canada Safeway, Limited
Canada Starch Company Limited
Canada Steamship Lines Limited
Canada Wire and Cable Company Limited
Canadair Limited
Canadian Acme Screw and Gear Limited
Canadian Bank of Commerce
Canadian Bank Note Company Limited

Canadian Bedding Company Limited, Edmonton
Canadian Blower and Forge Company Limited, Kitchener
Canadian Forest Products Limited, Vancouver
Canadian General Electric Company Limited
Canadian Ingersoll-Rand Company Limited
Canadian Kodak Company Limited
Canadian Marconi Company
Canadian Oil Companies Limited
Canadian Pacific Railway Company
Canadian Pratt and Whitney Aircraft Company Limited
Canadian Premier Life Insurance Company
Canadian Refractories Limited
Canadian Salt Company Limited
Canadian Sportswear Limited, Winnipeg
Canadian Superior Oil of California, Limited, Calgary
Canadian Utilities Limited, Edmonton
Canadian Wallpaper Manufacturing Limited
Canadian Western Natural Gas Company Limited, Edmonton
Canadian Westinghouse Company Limited
E. G. M. Cape and Company Limited
Cartier, Côté and Piette, Montreal
Century Geophysical Corporation of Canada, Calgary
Christie Grant's Limited, Edmonton
Confederation Life Association
Consumers' Gas Company, Toronto
Continental Can Company of Canada Limited
Continental Life Insurance Company
Cooperative Life Insurance Company, Regina
Crane Canada Limited
H. E. Crowell, Esq., Halifax
Crown Life Insurance Company
Cyanamid of Canada Limited
M. M. Dillon and Company, Limited
Dominion Bridge Company Limited
Dominion Chain Company Limited
Dominion Electrohome Industries Limited
Dominion Engineering Works Limited
Dominion Forge Limited
Dominion Foundries and Steel
Dominion Life Assurance Company
Dominion Rubber Company Limited
Dominion Stores Limited
Dow Chemical of Canada Limited
Dower Brothers Limited, Edmonton
Dufresne Engineering Company Limited, Montreal
Dupont Company of Canada Limited
John East Iron Works Limited, Saskatoon
Eaton Automotive Products Limited
The T. Eaton Company Limited
Eckler and Company Limited
Edmonton Furniture Limited
Edmonton Motors Limited
Edmonton Photo Supply Limited
Electroline Manufacturing Company Limited, Windsor
Empire Life Insurance Company
Equitable Life Insurance Company of Canada
H. M. E. Evans and Company, Edmonton
Excelsior Life Insurance Company
Falconbridge Nickel Mines Limited
Fashion Dress Shoppe Limited, Edmonton
Federal Grain Limited, Winnipeg
Federated Cooperatives Limited, Saskatoon
Fiberglass Canada Limited
Firestone Tire and Rubber Company of Canada, Limited
Fischer Bearings Manufacturing Limited
A. J. Freiman Limited, Ottawa
General Motors of Canada Limited
G. M. Gest, Contractors Limited
B. F. Goodrich Canada Limited
Goodyear Tire and Rubber Company of Canada, Limited
Great Atlantic and Pacific Tea Company Limited
G. L. Griffith and Sons Limited, Stratford

Helliwell, MacLachlan and Company, Vancouver
Home Oil Company Limited
Honeywell Controls Limited
Household Finance Corporation of Canada
Hudson's Bay Company
Hudson Bay Mining and Smelting Company Limited
A. T. Hurter, Esq., Montreal
Imperial Life Assurance Company of Canada
Imperial Tobacco Company of Canada Limited
Industrial Acceptance Corporation Limited
Industrial Life Insurance Company
International Business Machines Company Limited
International Harvester Company of Canada Limited
Interprovincial Pipe Line Company
Investors Syndicate of Canada Limited
Jenkins Brothers Limited, Montreal
Johnson Wire Works Limited, Montreal
Johnstone Walker Limited, Edmonton
K. C. S. Limited, Toronto
Harold Kline, Esq., Edmonton
Irving Kline Limited, Edmonton
Kraft Foods Limited
K. V. P. Company Limited
R. W. L. Laidlaw, Esq., Toronto
H. R. Lawson, Esq., Toronto
Lawson and Jones Limited, London
Lincoln Electric Company of Canada Limited
Loblaw Groceterias Company Limited
MacDonald Electric Limited, Kitchener
Madison Natural Gas Company Limited, Calgary
Maple Leaf Mills Limited
Marsland Engineering Limited, Kitchener
Massey-Ferguson Limited
Mathews Conveyer Company Limited, Port Hope
McArthur Ladies Wear, Edmonton
McCabe Grain Company Limited, Winnipeg
William M. Mercer Limited, Toronto
Metropolitan Life Insurance Company
Metropolitan Stores of Canada Limited
Mining Corporation of Canada Limited
Minnesota Mining and Manufacturing of Canada Limited
W. and A. Moir, Halifax
Molson's Brewery Limited
Monarch Life Assurance Company
Monsanto Canada Limited
Montreal City and District Savings Bank
Montreal Engineering Company Limited
Montreal Life Insurance Company
Henry Morgan and Company Limited
Mutual Life Assurance Company of Canada
National Cash Register Company of Canada Limited
National Grain Company Limited, Winnipeg
National Life Assurance Company of Canada
National Steel Car Corporation
A. D. Nesbitt, Esq., Montreal
Newhouse Wholesale Limited, Edmonton
Niagara Finance Company Limited
Carl and Nola Nickle Foundation
North American Life Assurance Company
Northern Life Assurance Company of Canada
Northwest Wholesale Furniture, Edmonton
Northwestern Utilities Limited, Edmonton
Norton Company of Canada Limited, Hamilton
The Ole Evinrude Foundation (Canada) Inc.
Ontario Paper Company Foundation
Otis Elevator Company, Hamilton
Page-Hersey Tubes, Limited
S. Panar, Edmonton
Parker Sportswear Company Limited, Edmonton
N. M. Paterson and Sons Limited, Winnipeg

Perini Limited
Philips Electronics Industries Limited
Pioneer Grain Company Limited, Winnipeg
Polymer Corporation Limited
K. A. Powell (Canada) Limited, Winnipeg
Premier Steel Mills Limited, Edmonton
Les Prévoyants du Canada
Promislow Brothers Limited, Regina
Provincial Bank of Canada
Prudential Assurance Company Limited of England
Pulp and Paper Mill Accessories Limited, Montreal
Quebec North Shore Paper Company
RCA Victor Company Limited
Remington Rand Limited
James Richardson and Sons Limited
Rolls-Royce of Canada Limited, Montreal
Mrs. Doris E. Rooney, Calgary
Royal Insurance Company Limited
Royal George Hotel, Edmonton
Royal Trust Company
Rule, Wynn and Rule, Edmonton
Sandwell and Company Limited, Vancouver
Sangamo Company Limited, Toronto
Savage Shoes Limited, Preston
Searle Grain Company Limited, Winnipeg
L. E. Shaw Limited, Halifax
Sheldons Engineering Limited, Galt
Shell Oil Company of Canada Limited
Shirriff-Horsey Corporation Limited
Siemens Edison Swan (Canada) Limited, Winnipeg
Silverwood Dairies Limited
Simpson-Sears Limited and
The Robert Simpson Company Limited
Henry Singer Limited, Edmonton
Howard Smith Paper Mills Limited
N. Smith Belting Works Limited, Toronto
Southam Company Limited
Southern Canada Power Company Limited
Sovereign Life Assurance Company of Canada
Sparton of Canada Limited, London
Sperry Gyroscope Company of Canada Limited
Spruce Falls Power and Paper Company Limited
Standard Life Assurance Company
Standard Oil Company of British Columbia Limited
Sterling Printers, Edmonton
Stone and Webster Canada Limited
Supertest Petroleum Corporation Limited, London
Taylor, Pearson and Carson (Canada) Limited
Texaco Canada Limited
B. B. Torchinsky and Associates Limited, Saskatoon
Toronto-Dominion Bank
Traders Finance Corporation Limited
Trans Mountain Oil Pipe Line Company
Trudeau's Limited, Edmonton
United-Carr Fastener Company of Canada Limited, Hamilton
United Geophysical Company of Canada, Calgary
Vancouver Sun
Ventures Limited, Toronto
Waterloo Motors, Edmonton
Weisler's Limited, Edmonton
W. C. Wells, Construction Company Limited, Saskatoon
Westclox Canada Limited, Peterborough
Western Life Assurance Company
Winnipeg Free Press Company Limited
Winspear, Hamilton and Anderson, Edmonton
The Honourable Robert H. Winters, Toronto
Woodward Stores Limited, Vancouver
Max Wyman, Esq., University of Alberta

This list might have contained also almost all of the universities of Canada; they have supported the Congress by sending or helping to send members of their staffs to our meetings and in many cases by contributing directly to the support of the *Canadian Journal of Mathematics* rather than contributing directly to our treasury. Their contribution is of the greatest importance and is acknowledged with gratitude. Gratitude is also expressed to many individuals who contributed to the success of the Congress and Seminar of 1961, either financially or otherwise.

The National Research Council gave not only generous financial support but also the lively interest of its members and particularly of its late president, Dr. E. W. R. Steacie.

We record here our thanks to the Canada Council not only for $2,000 towards the cost of publication of these Proceedings and $2,000 towards the expenses of the research lecturers, but also for the interest shown in the development of mathematics. In particular, we are grateful for the interest shown by its director, Dr. A. W. Trueman, in the efforts of the Congress to strengthen the contribution of mathematics to the cultural life of Canada.

This was the second meeting of the Congress in Montreal. The first one was held in 1945 at McGill University when the organization was founded. We are grateful to the rector of the University of Montreal, Monseigneur Lussier, for his co-operation in extending to the Congress the use of the facilities of the University and to the staff for their untiring assistance. Many were involved in the advance preparations but special mention should be made of the members of the Department of Mathematics who did so much to make the Seminar and Congress of 1961 another outstanding success.

L. F. S. RITCEY
Executive Director

PRESIDENTIAL ADDRESS

DERIVATIVES AND INTEGRALS WITH RESPECT TO A BASE FUNCTION

R. L. JEFFERY

1. Introduction. This study arose out of work started by Lusin **(1)** which led to the following theorems:

Let $F(x)$ and $G(x)$ be two functions on $[a, b]$ which are continuous and generalized absolutely continuous, (ACG). If $F'(x) = G'(x)$, except for a set of measure zero, then $F(x)$ and $G(x)$ differ by a constant.

If $F(x)$ is continuous and ACG on $[a, b]$ and if $F'(x) = f(x)$ except for a set of zero measure then

$$F(x) - F(a) = \int_a^x f(t)dt$$

where the integration is in the generalized Denjoy sense.

In this address corresponding results are presented when the measure function and integration are in terms of a base function ω which is non-decreasing.

The study was carried out during the 1961 session of the Summer Research Institute of the Canadian Mathematical Congress and the 1962 session of the Summer Research Institute of the Australian Mathematical Society. The outcome was a constructive approach to a generalized integral for which the base function is first of bounded variation and then of generalized bounded variation in the restricted sense. The latter results will be published elsewhere. Fundamental results in the same direction have been obtained by Ward **(2)** and by Henstock **(3)**. Their work is based on the descriptive approach of Perron.

In both studies there is involved what might be called a point by point examination of the functions which enter. For this reason it was found desirable to give, in terms suitable for the purpose in hand, the definition of a measure function and to state its properties which will be used.

1.2. *Functions in class* $\mathfrak{F}$. A function F in class $\mathfrak{F}$ has the following properties. F is defined on $(-\infty, \infty) = \mathfrak{J}$ except at the points of a denumerable set N. For $x \in \mathfrak{J}$ $F(x+)$ and $F(x-)$ both exist and are finite where the limits are taken over $\mathfrak{J} - N$. For $x \in \mathfrak{J} - N$, $F(x+) = F(x-) = F(x)$.

1.3. *An outer measure function relative to a function* $\omega \in \mathfrak{F}$ *which is non-decreasing and bounded.* Let $\omega(x) \in \mathfrak{F}$ be bounded on $\mathfrak{J} - N$ and be such that if $x_1, x_2 \in \mathfrak{J} - N$, $x_1 > x_2$, then $\omega(x_1) \geqslant \omega(x_2)$.

For an open interval (a, b) let $\mu^*(a, b) = \omega(b-) - \omega(a+)$. For any set $A \in \mathfrak{J}$ let

$$\mu^*(A) = \inf \sum \mu^*(x_i, x_i')$$

where (x_i, x_i') is a set of open intervals covering A. If θ is the empty set $\mu^*(\theta) = 0$. $\mu^*(A)$ is then an outer measure function in the sense of Monroe (**2**, chapter II).

In what follows lower case Greek letters denote open sets and for any set A, $\tilde{A}$ denotes the complement of A, $\tilde{A} = \mathfrak{J} - A$.

If A is any set and $\alpha \supset A$ then $\alpha = \cup \alpha_i$, $\alpha_i \cap \alpha_j = 0$, $i \neq j$, $\mu^*(A) \leqslant \mu^*(\alpha_i)$. Results I–VI follow from (**4**, chapter II, **5**).

I. $\mu^*(A) \leqslant \sum \mu(\alpha_i) \leqslant \lim_{a \to -\infty,\, b \to \infty} \{\omega(b-) - \omega(a+)\}$

$$= \mu^*(\mathfrak{J}) < \infty.$$

II. If $A \subset B$ then $\mu^*(B) \leqslant \mu^*(A)$.

III. For any sets A, B the relation $\mu^*(A \cup B) \leqslant \mu^*(A) + \mu^*(B)$ holds.

Two sets A and B are ω-separated if for $\epsilon > 0$, ϵ arbitrary, there is $\alpha \supset A$, $\beta \supset B$ with $\mu^*(\alpha \cap \beta) < \epsilon$. If the set A is such that A and $\tilde{A}$ are ω-separated then A is ω-measurable. The measure of A is $\mu^*(A)$ and is denoted by $\mu(A)$.

IV. Open sets and closed sets are ω-measurable.

V. If $A = \cup A_i$ where A_i is ω-measurable and if A_i, A_j are ω-separated or if $A_i \cap A_j = 0$, $i \neq j$, then the set A is ω-measurable and $\mu(A) = \sum \mu(A_i) < \infty$.

VI. If $A = \cup A_i$ where $A_n \supset A_{n-1}$ then

$$\mu^*(A) = \lim_{n \to \infty} \mu^*(A_n).$$

2. Derivatives with respect to a base function. This section deals with the derived numbers of functions $F \in \mathfrak{F}$ with respect to a base function ω, and with functions in $\mathfrak{F}$ which are absolutely continuous, and those which are generalized absolutely continuous, with respect to a base function. For any $x \in \mathfrak{J}$, for $x + h \in \mathfrak{J} - N$, and for $\omega(x+h) - \omega(x\pm) \neq 0$, set

$$\psi(x, h) = \frac{F(x+h) - F(x-)}{\omega(x+h) - \omega(x-)} \qquad h > 0$$

$$\psi(x, h) = \frac{F(x+h) - F(x+)}{\omega(x+h) - \omega(x+)}, \qquad h < 0$$

$$= 0, \ \omega(x+h) - \omega(x\pm) = 0.$$

The derived numbers of F with respect to ω are the upper and lower limits as $h \to 0$ of $\psi(x, h)$. If the limit exists it is the derivative of F with respect to ω, $D_\omega F$. At points of discontinuity of ω, $D_\omega F$ exists even though one or both of F, ω may not be defined at such points.

For the proofs of the theorems which follow we shall need

LEMMA 2.1. *Let E be a set on an interval (a, b). Let each point x of E be the left end-point of a set of intervals $[x, x + h_i]$, $h_i > 0$, $h_i \to 0$. Let V denote the set of intervals thus associated with the set E. If $\epsilon > 0$ is given there is then a mutually exclusive finite set of closed intervals $v_1, v_2, \ldots, v_p$, for which*

$$\mu(\cup v_i) < \mu^*(E) + \epsilon, \qquad \mu^*(\cup E \cap v_i) > \mu^*(E) - \epsilon.$$

A similar lemma is given in (**7**, p. 618), where the intervals are non-overlapping rather than mutually exclusive.

Let $\alpha \supset E$ with $\mu(\alpha) < \mu^*(E) + \epsilon$. Delete from V all intervals not on α. Then $\alpha = \cup \alpha_i$, $\alpha_i \cap \alpha_j = 0$, $i \neq j$. Let

$$\alpha^n = \bigcup_i^n \alpha_i$$

with n sufficiently great to insure that for ϵ arbitrary, but fixed,

$$\mu^*(E \cap \alpha^n) > \mu^*(E) - \epsilon.$$

Let (a_i, b_i) be an interval of α^n. Let $\delta_1 > \delta_2 > \ldots$ be a sequence of positive numbers with $\delta_k \to 0$ and let $E_k \subset E \cap (a_i, b_i)$ be such that if $x \in E_k \subset E \cap (a_i, b_i)$ there is at least one interval $[x, x+h_k]$ of V which is on (a_i, b_i) with $h_k > \delta_k$ and with $x + h_k \in (a_i, b_i)$. By VI, § 1.3, $\mu^*(E_k) \to \mu^* E \cap (a_i, b_i)$ as $\delta_k \to 0$. Each point $x \in E \cap (a_i, b_i)$ is in E_k for δ_k sufficiently near zero or for k sufficiently great.

Now take k so great that

$$\mu^*(E_k) > \mu^*[E \cap (a_i, b_i)] - \frac{\epsilon}{2n},$$

n the index of α^n. Let $x_0 = \inf E_k$. If $x_0 \in E_k$, let

$$v_1 = [x_0, x_0 + h_0], \; h_0 = h_{x_0} > \delta_k.$$

If $x_0 \notin E_k$ take $x_1 > x_0$, $x_1 \in E_k$, and such that

$$\mu^*[E_k \cap (x_0, x_1)] < \epsilon_1 \tag{1}$$

where ϵ_1 is the first of a sequence $\epsilon_1, \epsilon_2, \ldots$, for which $\sum \epsilon_i < \epsilon/2n$. In this case let $v_1 = [x_1, x_1 + h_1]$, $h_1 > \delta_k$.

Take $x_2 \in E_k$, $x_2 > x_1 + h_1$ and so that

$$\mu^*[E_k \cap (x_1 + h_1, x_2)] < \epsilon_2, \tag{2}$$

and take $v_2 = [x_2, x_2 + h_2]$, $h_2 > \delta_k$. This procedure can be continued, and because $\delta_k > 0$ is fixed and because $x + h_x \in (a_i, b_i)$ for $x \in E_k$, it follows that an integer p is reached for which $\cup v_i$, $i = 1, \ldots, p$, contains all of E_k on $[a, b]$ except the set $\cup E_k \cap (x_i + h_i, x_{i+1})$. It then follows from relations such as (1) and (2) that

$$\mu^*[\cup (E_k \cap v_j)] > \mu^*[E_k \cap (a_i, b_i)] - \sum \epsilon_i$$

$$\mu^*[\cup E \cap v_j] > \mu^*[E_k \cap (a_i, b_i)] - \frac{\epsilon}{n}. \tag{3}$$

Then, because $E_k \subset E$, it follows that

$$\mu^*(\cup E \cap v_j) > \mu^*[E \cap (a_i, b_i)] - \frac{\epsilon}{n}.$$

The closed intervals v_j are mutually exclusive.

If this construction is carried out for each of the intervals of the set α^n then the total set of intervals v_i thus obtained will be finite in number, will be mutually exclusive, and, because $v_i \in \alpha^n \subset \alpha$, it follows that

$$\mu(\cup v_i) < \mu^*(E) + \epsilon.$$

Also, because of (3), the total set v_i is such that

$$\mu^*[\cup E \cap v_i] > \mu^*(E) - \epsilon.$$

THEOREM 2.1. *If F is non-decreasing, $F \in \mathfrak{F}$, then $D_\omega F$ exists except for a set of ω-measure zero.*

Let E_δ be the set for which $\bar{D}F^+ - \underline{D}F^+ > \delta$ and suppose that for δ sufficiently near zero $\mu^*(E_\delta) > 0$. Let y_i be a dense denumerable set on $-\infty < y < \infty$ and let

$$E_i = [x: x \in E_\delta, y_i - \tfrac{1}{4}\delta < \bar{D}_\omega F(x)^+ < y_i + \tfrac{1}{4}\delta].$$

If $x \in E_\delta$, then $x \in E_i$ for at least one value of i. Also, because $\{E_i\}$ is a denumerable sequence and because $\mu^*(E_\delta) > 0$, it follows that $\mu^*(E_i) > 0$ for at least one value i_0 of i. Let $E_{i_0} = E_0$, $y_{i_0} = y_0$. Then for $x \in E_0$

$$(1)\quad d = y_0 - \tfrac{1}{4}\delta < \bar{D}F^+, \qquad \underline{D}F^+ < y_0 - \tfrac{3}{4}\delta = c, \qquad d - c > \tfrac{1}{2}\delta.$$

Also for $x \in E_0$ we have

$$(2)\qquad \frac{F(x + h_i) - F(x-)}{\omega(x + h_i) - \omega(x-)} < c$$

for a sequence $\{h_i\}$, $h_i > 0$, $h_i \to 0$, $x + h_i \in \mathfrak{J} - N$. Lemma 2.1 can now be used to select from the intervals associated with E_0 in (2) a mutually exclusive set of closed intervals $v_1^0, v_2^0, \ldots, v_n^0$, $v_i^0 = [x_i, x_i']$, $x_i \in E_0$, $x_i' \in \mathfrak{J} - N$, and

$$(3)\qquad \frac{F(x_i') - F(x_i-)}{\omega(x_i') - \omega(x_i-)} < c$$

and for which, if ϵ is an arbitrary positive number,

$$(4)\quad \mu(\cup v_i^0) < \mu^*(E_0) + \epsilon, \qquad \mu^*[\cup (E_0 \cap v_i^0)] > \mu^* E_0 - \epsilon.$$

Because $F(x)$ and $\omega(x)$ are continuous at x_i' and because $F(x_i-)$ and $\omega(x_i-)$ exist there is an open interval (ξ_i, ξ_i') with $\xi_i < x_i < x_i' < \xi_i'$, with $\xi_i, \xi_i' \in \mathfrak{J} - N$ and with

$$(5)\qquad \frac{F(\xi_i') - F(\xi_i)}{\omega(\xi_i') - \omega(\xi_i)} < c.$$

Furthermore, because $\{v_i^0\}$ is a mutually exclusive set of closed intervals, ξ_i, ξ_i' can be so chosen that the intervals (ξ_i, ξ_i') are mutually exclusive, and so that

$$\mu[\cup (\xi_i, \xi_i')] < \mu^*(E_0) + 2\epsilon. \tag{6}$$

From (5) and (6) we then have, because $F(x)$ is non-decreasing,

$$\sum\{F(\xi_i') - F(\xi_i)\} < c\{\mu^*(E_0) + 2\epsilon\}. \tag{7}$$

Let

$$E^0 = E\{x: x \in \cup(E_0 \cap v_i^0)\}.$$

Then, because $E^0 \subset E_0$ and $E^0 \subset \cup(\xi_i, \xi_i')$, we have from the first part of (1), for $x \in E^0$, a sequence $\{h_i\}$, $h_i > 0$, $h_i \to 0$, $x + h_i \in \mathfrak{J} - N$, $(x, x + h_i)$ on an interval (ξ_j, ξ_j'), and with

$$\frac{F(x + h_i) - F(x-)}{\omega(x + h_i) - \omega(x-)} > d. \tag{8}$$

Lemma 2.1 can now be used to select from the set of intervals associated with E^0 by (8) a mutually exclusive set $v_1', v_2', \ldots, v_p'$, $v_i' = [x_i, x_i']$, for which

$$\frac{F(x_i') - F(x_i-)}{\omega(x_i') - \omega(x_i-)} > d, \tag{9}$$

$$\mu(\cup v_i') < \mu^*(E^0) + \epsilon, \qquad \mu^*[E^0 \cap (\cup v_i')] > \mu^*E^0 - \epsilon. \tag{10}$$

Then relations (9) and (10), with the last part of (4), give

$$\sum\{F(x_i') - F(x_i-)\} > d\{\mu^*(E_0) - 2\epsilon\}. \tag{11}$$

But because each (x_i, x_i') is on an interval (ξ_j, ξ_j') and because F is non-decreasing we have, using (7),

$$\sum\{F(x_i') - F(x_i)\} \leqslant \sum\{F(\xi_i') - F(\xi_i)\} < c\{\mu^*(E_0) + 2\epsilon\}. \tag{12}$$

Then, because $d - c > \frac{1}{2}\delta$ and ϵ is arbitrary independently of δ, it follows that (11) and (12) are contradictory. We conclude, therefore, that $\bar{D}F^+ = \underline{D}F^+ = DF^+$ except for a set of ω-measure zero.

The proof of Lemma 2.1 can obviously be altered to conform to the case in which each point of a set E is the right end-point of a sequence of intervals $[x + h, x]$, $h < 0$. With Lemma 2.1 altered in this way it can be used, as in the foregoing, to show that the left derivative of F with respect to ω exists except for a set of ω-measure zero. The derivative exists at points of discontinuity of ω, and it can then be shown, as in (**5**, p. 123) that the left derivative and the right derivative are equal except for a denumerable set at each of which ω is continuous, hence a set which has ω-measure zero. Obviously the left and right derivatives are equal at a point of discontinuity of ω. This completes the proof of Theorem 2.1.

THEOREM 2.2. *If $D_\omega F$ exists on an ω-measurable set E with $\mu(E) > 0$ then $D_\omega F$ is ω-measurable on E.*

Let a be a real number and let

$$E^a = E[x: x \in E, D_\omega F < a], \qquad E_a = E[x: x \in E, D_\omega F \geqslant a].$$

If $D_\omega F$ is not measurable on E then for some a the sets E^a, E_a are not ω-separated. If such is the case, for $c > 0$, let

$$E^{ca} = E[x: x \in E^a, D_\omega F < a - c].$$

If $c_1 < c_2$, $E^{c_1 a} \supset E^{c_2 a}$. Hence by VI, § 1.3, $\mu^*(E^{ca}) \to \mu^*(E^a)$ as $c \to 0$. It then follows that for c sufficiently near zero the sets E^{ca} and E_a are not ω-separated, and $\mu^*(E^{ca}) > 0$. Now for $\delta > 0$ let $E_\delta{}^{ca}$ be such that for $x \in E_\delta{}^{ca} \subset E^{ca}$

$$\frac{F(x+h) - F(x-)}{\omega(x+h) - \omega(x-)} < a - c$$

$0 < h < \delta$, $x + h \in \mathfrak{J} - N$. Then for $\delta_1 < \delta_2$

$$E_{\delta_1}^{ca} \supset E_{\delta_2}^{ca}.$$

Hence, by VI, § 1.3, $\mu^*(E_\delta{}^{ca}) \to \mu^*(E^{ca})$ as $\delta \to 0$. Then for δ sufficiently near zero $E_\delta{}^{ca}$ and E_a are not ω-separated. It then follows that there are points of E_a which are limit points of $E_\delta{}^{ca}$. Let $x \in E_a$ be a limit point of $E_\delta{}^{ca}$ on the left and fix h with $0 < h < \delta$, $x + h \in \mathfrak{J} - N$ so that

$$\text{(1)} \qquad \frac{F(x+h) - F(x-)}{\omega(x+h) - \omega(x-)} > a - \frac{c}{2}.$$

There is then a sequence $x_i \in E_\delta{}^{ca}$ $x_i \to x$ on the left $x_i + h_i = x + h$, $0 < h_i < \delta$, and with

$$\text{(2)} \qquad \frac{F(x_i + h_i) - F(x_i-)}{\omega(x_i + h_i) - \omega(x_i-)} = \frac{F(x+h) - F(x_i-)}{\omega(x+h) - \omega(x_i-)} < a - c.$$

As $x_i \to x$ the left side of (2) tends to the left side of (1) which leads to a contradiction. Hence we can conclude that E^a and E_a are ω-separated.

Because $E^a \supset E$ where E is ω-measurable it follows that E^a and $\tilde{E}$ are ω-separated. Then because $\tilde{E}^a = E_a + \tilde{E}$ and because E^a and E_a are ω-separated it follows that E^a and $\tilde{E}^a$ are ω-separated which means that E^a is ω-measurable.

3. Absolute continuity and generalized absolute continuity relative to ω. Let F be a function, $F \in \mathfrak{F}$, E a set $E \in \mathfrak{J}$. If to $\epsilon > 0$ there corresponds $\delta > 0$ which is such that for every non-overlapping set of intervals (x_i, x_i') with $x_i, x_i' \in E$ and $\sum \mu^*[E \cap (x_i, x_i')] < \delta$ the relation

$$\sum |F(x_i'+) - F(x_i-)| < \epsilon$$

holds, then F is absolutely continuous over E relative to ω, $AC-\omega$.

If (a, b) is an interval and $(a, b) = \cup E_i$, E_i closed; and if F is $AC - \omega$ over each E_i, then F is generalized absolutely continuous relative to ω on (a, b), ACG on (a, b).

THEOREM 3.1. *If $F \in \mathfrak{F}$ is $AC - \omega$ on (a, b) and if $D_\omega F = 0$ except for a set of ω-measure zero then F is constant on $(a, b) - N$.*

For the proof of Theorem 3.1 we note first that $F(x_0+) = F(x_0-)$ if x_0 is a point of discontinuity of ω. For otherwise at x_0, $D_\omega F > 0$ and $\mu(x_0) > 0$. Secondly, if x_0 is a point of continuity of ω then $F(x_0+) = F(x_0-)$, for otherwise F is not $AC - \omega$ on (a, b). Thus if at points $x \in N$ at which F is not defined it is agreed that $F(x) = F(x+) = F(x-)$ then $F(x)$ is continuous and $AC - \omega$ on (a, b). It does not necessarily follow that F is AC on (a, b).

Let E be the set on (a, b) which is such that $D_\omega F = 0$ for $x \in E$. Then, because $\mu(\tilde{E}) = 0$, the set E is ω-measurable and $\mu(E) = \mu(a, b)$. Let $\epsilon > 0$ be given. For $x \in E$ there is a sequence $x + h_i$, $h_i \to 0$, $h_i > 0$, $x + h_i \in \mathfrak{J} - N$, for which

$$\frac{F(x + h_i) - F(x-)}{\omega(x + h_i) - \omega(x-)} < \epsilon. \tag{1}$$

This associates with $x \in E$ a set of intervals of length arbitrarily near zero for which (1) holds. It then follows from Lemma 2.1 that there is a finite set of mutually exclusive closed intervals $v_i = (x_i, x_i')$ for which

$$\frac{|F(x_i') - F(x_i-)|}{\omega(x_i') - \omega(x_i-)} < \epsilon \tag{2}$$

and for which

$$(3) \qquad \sum\mu(v_i) > \mu(a, b) - \epsilon, \qquad \sum\mu(\tilde{v}_j) < \epsilon,$$

where $\tilde{v}_j$ are the intervals complementary to the closed set v_i. The point x_i' is a point of continuity of both F and ω. Furthermore, for any x there is a sequence x_i with $x_i < x$, $x_i \to x$, x_i points of continuity of ω and $\omega(x_i) \to \omega(x-)$ as $i \to \infty$. Hence, since the intervals v_i are closed and mutually exclusive, x_i in (2) can be moved to the left so that the set $[x_i, x_i']$ is again mutually exclusive, so that x_i is a point of continuity of ω and so that (3) holds and

$$(4) \qquad \frac{|F(x_i') - F(x_i)|}{\omega(x_i') - \omega(x_i)} < \epsilon.$$

Furthermore, if η were assigned and arbitrary ϵ could be taken so that the second relation in (3), with $\tilde{v}_j = (\xi_j, \xi_j')$, gives

$$\sum |F(\xi_j') - F(\xi_j)| < \eta$$

because F is $AC - \omega$ on (a, b). It follows from (2) that for the intervals $v_i = [x_i, x_i']$

$$\sum |F(x_i') - \sum F(x_i)| < \epsilon\mu(a, b).$$

Then

$$|F(b-) - F(a+)| = |\sum \{F(x_i') - F(x_i)\}$$
$$+ \sum \{F(\xi_j') - F(\xi_j)\}| \leqslant \epsilon\mu(a, b) + \eta$$

where ϵ and η are arbitrary. It then follows that $F(b-) = F(a+)$. For x with $a < x < b$ the same procedure can be used to show that $F(x-) = F(a+)$, or since $F(x)$ is continuous on (a, b), that $F(x) = F(a+)$. Hence $F(x)$ is constant on (a, b), and the original $F(x)$ is constant on $(a, b) - N$.

THEOREM 3.2. *Let* $F(x) \in \mathfrak{F}$ *be* ACG $- \omega$ *on the interval* (a, b) *and let* $D_\omega F = 0$ *except for a set of* ω*-measure zero. Then* F *is constant on* $(a, b) - N$.

The interval (a, b) is covered by $\cup E_i$ where E_i is closed and F is $AC - \omega$ over each E_i. It follows as in Theorem 3.1 that if x_0 is a point of discontinuity of ω then $F(x_0+) = F(x_0-)$, for otherwise $D_\omega F \neq 0$ at x_0 with $\mu(x_0) > 0$. If x_0 is a point of continuity of ω

then $x_0 \in E_i$ for at least one value of i and if $F(x_0+) \neq F(x_0-)$ then F is not $AC - \omega$ over E_i. Then, as in Theorem 3.1, $F(x+) = F(x-)$ for $x \in (a, b)$. If it is agreed that $F(x) = F(x+) = F(x-)$ then F becomes continuous on (a, b).

Let $[a', b']$, $a < a' < b' < b$, be a closed interval on (a, b) with $[a', b']$ points of continuity of ω. Then $[a', b'] = \bigcup\{[a', b'] \cap E_i\} = \bigcup\mathfrak{E}_i$, $\mathfrak{E}_i$ is closed, and F is $AC - \omega$ on $\mathfrak{E}_i$. By Baire's theorem there is an integer n and an interval (l, m) on $[a', b']$ for which, if $\mathfrak{E} = \bigcup\mathfrak{E}_i$,

$$[a', b'] \cap (l, m) = \mathfrak{E}^n(l, m) = (l, m).$$

Then F is $AC - \omega$ on (l, m), and it follows from Theorem 3.1 that F is constant on (l, m).

Thus there are points $x \in [a', b']$ such that about each such point there are intervals on which F is constant. Let E be the points on $[a', b']$ which is such that if $x \in E$ there then is no interval containing x on its interior on which F is constant.

The set E is closed. Let (a_i, b_i) be an interval of $\tilde{E} \cap [a', b']$ and let $[l', m']$ be a closed interval with $a_i < l' < m' < b_i$. Each point $x \in [l', m']$ is on an interval α_x on which F is constant. By the Heine-Borel theorem there is a finite set of intervals covering $[l', m']$ on each of which F is constant. It then follows from the continuity of F that F is constant on $[l', m']$. Then since l', m' may be taken arbitrarily near a_i and b_i respectively it follows that F is constant on (a_i, b_i), and, since F is continuous, on $[a_i, b_i]$.

The set E is closed and is covered by the sequence of closed sets $E \cap E_i$. It again follows from Baire's theorem that there is an integer n and an interval (l, m) such that $E \cap (l, m) = E_n \cap (l, m) = e$. Then F is $AC - \omega$ over e. Let $\epsilon > 0$ be given. Fix $\delta > 0$ so that if (x_i, x_i') is any set of non-overlapping intervals on (l, m) with $x_i, x_i' \in e$ with $\sum\mu\{e \cap [x_i, x_i']\} < \delta$ then

$$\sum|F(x_i') - F(x_i)| < \epsilon. \tag{1}$$

Now let (ξ_j, ξ_j') be *any* set of non-overlapping intervals on (l, m) with $\sum\mu[\xi_j, \xi_j'] < \delta$. Then $\mu\{\bigcup(e \cap [\xi_j, \xi_j'])\} < \delta$. If (ξ_j, ξ_j') contains no points of e then F is constant on $[\xi_j, \xi_j']$ and $F(\xi_j') - F(\xi_j) = 0$. If (ξ_j, ξ_j') contains points of e and $\xi_j \notin e$ there is a first point $\bar{\xi}_j \in e$ to the right of ξ_j and $F(\bar{\xi}_j) = F(\xi_j)$. Similarly if $\xi_j' \notin e$ there is a first point $\bar{\xi}_{j'}'$ to the left of $\xi_{j'}'$ with $F(\bar{\xi}_j') = F(\bar{\xi}_j')$ and $F(\bar{\xi}_j') - F(\xi_j) = F(\bar{\xi}_j') - F[\bar{\xi}_j]$. It could be that $\bar{\xi}_{j'}' = \bar{\xi}_j$. If $\xi_j, \xi_{j'}' \in e$, denote $\xi_{j'}$ by $\bar{\xi}_j$, $\xi_{j'}'$ by $\bar{\xi}_{j'}'$. It now follows from this and (1), because $\bar{\xi}_j, \bar{\xi}_j' \in e$, that

$$\sum|F(\xi_j') - F(\xi_j)| = \sum|F(\bar{\xi}_j') - F(\bar{\xi}_j)| < \epsilon.$$

It may, therefore, be concluded that F is $AC - \omega$ on (l, m), and it follows from Theorem 3.1 that F is constant on (l, m). But this permits the conclusion that E is empty.

It now follows that every point x of $[a', b']$ is on an interval α_x on which F is constant and, as in the case of the interval $[l', m']$ above, the Hirne–Borel theorem may be applied to show that F is constant on $[a', b']$. Because $a' > a$ and $b' < b$ may be taken arbitrarily near a and b respectively it follows that F is constant on (a, b). Hence the original F is constant on $(a, b) - N$.

THEOREM 3.3. *Let F and G be two functions in class $\mathfrak{F}$ each of which is $ACG - \omega$ on (a, b) and which are such that $D_\omega F = D_\omega G$ except for a set of ω-measure zero. Then the sets $[x, F(x)]$, $[x, G(x)]$, $x \notin N$, are identical or one is a translation parallel to the y-axis of the other.*

If $x_0 \in (a, b)$ is a point of discontinuity of ω then at x_0 both $D_\omega F$ and $D_\omega G$ exist and are equal. For otherwise it could not be said that $D_\omega F = D_\omega G$ except on a set of ω-measure zero. Again, because $D_\omega F = D_\omega G$ at a point x_0 of discontinuity of ω, the jumps of F and G at such a point must be the same and in the same direction. Now set

$$H(x) = F(x) - G(x). \tag{1}$$

Then $D_\omega H = 0$ except at a set of ω-measure zero. Furthermore

$$(a, b) = \cup E_i, \qquad (a, b) = \cup E_i'$$

with each E_i and each E_i' closed and with F $AC - \omega$ over each E_i and G $AC - \omega$ over each E_i'. It then follows that H is $AC - \omega$ over each of the closed sets $E_i \cap E_j'$, $i, j = 1, 2, \ldots$. It then follows as in Theorem 3.2 that if $H(x)$ is defined by $H(x) = F(x+) = F(x-)$ then $H(x)$ is continuous and $ACG - \omega$ on (a, b) and is, therefore, constant by Theorem 3.2. The truth of the theorem to be proved now follows from relation (1).

THEOREM 3.4. *Let* $F \in \mathfrak{F}$ *be defined on* $(a, b) - N$ *where* N *is the set of discontinuities of* ω. *Let* F *be* $ACG - \omega$ *on* (a, b), *and let* $D_\omega F = f$ *exist and be finite except at a set of* ω*-measure zero. Then if* f *is given it is possible to determine* $F(b-) - F(a+)$ *in a denumerable set of operations on* f.

This theorem is given in **(7**, p. 623**)**, but there is an error in the proof. The point x_i' does not necessarily belong to e.

Because F is $ACG - \omega$ on (a, b) it follows that $(a, b) = \cup E_i$, E_i closed and F is $AC - \omega$ over each E_i. If x_0 is a point of discontinuity of ω then $D_\omega F$ exists at x_0. For otherwise $D_\omega F$ fails to exist at a set of ω-measure greater than zero.

On the other hand, if x_0 is a point of continuity of ω then x_0 belongs to at least one of the sets E_i and unless $F(x_0+) = F(x_0-)$ the function F is not $AC - \omega$ over E_i. We thus have $F(x)$ continuous at the points of continuity of ω. Also it follows that F is constant on intervals on which ω is constant. For otherwise F would not be $ACG - \omega$ on (a, b).

Let E be any closed set on $[a', b']$ where $[a', b']$ is a closed interval with $a < a' < b' < b$ and with a', b' points of continuity of ω. Then $E = \cup EE_i$, and $EE_i = \mathfrak{E}_i$ is closed. Hence, by Baire's theorem, there is an interval (l, m) and an integer n such that $E(l, m) = \mathfrak{E}_n(l, m) = \mathfrak{E}$, and $F(x)$ is $AC - \omega$ over $\mathfrak{E}$. Then there is a number $M > 0$ such that if (x_i, x_i') is any set of non-overlapping intervals with end-points in $\mathfrak{E}$,

$$\sum |F(x_i'+) - F(x_i-)| < M. \tag{1}$$

Now

$$|F(x_i'-) - F(x_i+)| = |F(x_i'+) - F(x_i-) - \{F(x_i'+) - F(x_i'-)\} - \{F(x_i+) - F(x_i-)\}|$$

Hence

$$\sum |F(x_i'-) - F(x_i+)| < 3M. \tag{2}$$

Let $[l', m']$ be an interval with $l < l' < m' < n$, with l', m' points of continuity of ω, and with

(3) $$\mu[\mathfrak{E} \cap [l', m'] > \mu[\mathfrak{E} \cap (l, m)] - \epsilon,$$

where ϵ is arbitrary. The set $\mathfrak{E}[l', m'] = \mathfrak{E}'$ is closed, and by Theorem 2.2 $D_\omega F$ is ω-measurable over $\mathfrak{E}'$. We show that $|f| = |D_\omega F|$ is Lebesgue–Stieltjes integrable with respect to ω over $\mathfrak{E}'$. Let (l_{i-1}, l_i) be a subdivision of $(-a, \infty)$, $a > 0$ but arbitrarily near zero, $l_0 = -a$, and let

$$e_i = E[x: x \in \mathfrak{E}', l_{i-1} < |f| \leqslant l_i].$$

Suppose that $\sum l_{i-1}\mu(e_i)$ diverges. Then there is n such that

(4) $$\sum_{i=1}^{n} l_{i-1}\mu(e_i) > 2M.$$

For $x \in e_0$ there is δ_x such that

(5) $$\frac{|F(x+h) - F(x-)|}{\omega(x+h) - \omega(x-)} > l_{i-1},$$

$0 < h < \delta_x$, $x + h \in \mathfrak{J} - N$.

Lemma 2.1 can now be used to select from the set of intervals $[x, x+h]$ thus associated with the points $x \in e_0$ a finite mutually exclusive set of intervals $v_1{}^0, v_2{}^0, \ldots, v_p{}^0$ for which

$$\mu(\bigcup v_j^0) < \mu(e_0) + \epsilon_0 \qquad \mu[e_0 \cap (\bigcup v_j^0)] > \mu(e_0) - \epsilon_0$$

where ϵ_0 is arbitrary. Because e_0 and e_1 are ω-measurable it follows that if $\bigcup \bar{v}_j$ is the set of intervals complementary to $\bigcup v_j{}^0$ then

(6) $$\mu[\bigcup(e_1 \cap \bar{v}_j)] > \mu(e_1) - \epsilon_0.$$

It is then possible to use Lemma 2.1 to determine a finite set of mutually exclusive intervals $v_1{}^1, v_1{}^1, \ldots, v_q{}^1$ with each $v_j{}^1$ on an interval of the set $\bar{v}_j$ and with

$$\mu(\bigcup v_j^1) < \mu[\bigcup (e_1 \cap \bar{v}_j)] + \epsilon_1,$$

$$\mu(\bigcup e_1 \cap v_j^1) > \mu(\bigcup e_1 \cap \bar{v}_j) - \epsilon_1 > \mu(e_1) - \epsilon_0 - \epsilon_1.$$

Since each closed interval of $v_j{}^1$ is on an interval of the set $\bigcup \bar{v}_j$ it follows that the intervals of the set $(\bigcup v_j{}^0) \cup (\bigcup v_j{}^1)$ are mutually exclusive.

Let $[x_{0j}, x_{0j}']$, $[x_{1j}, x_{1j}']$ be the intervals of the respective sets of v_j^0, v_j^1. Then, because x_{0j}', x_{1j}' are points of continuity of F and of ω, it follows from (5) that

$$\begin{aligned}\sum |F(x_{0j}') - F(x_{0j}-)| &> \sum \{\omega(x_{0j}') - \omega(x_{0j}-)\}l_0 \\ &> l_0[\mu(e_0)-\epsilon_0]\end{aligned}$$

$$\sum |F(x_{1j}') - F(x_{1j}-)| > l_1[\mu(e_1) - \epsilon_0 - \epsilon_1].$$

This procedure can be continued to arrive at a mutually exclusive set of closed intervals $[x_k, x_k']$ with $x_k \in \mathfrak{E}'$, x_k' a point of continuity of F and with

$$\begin{aligned}\sum |F(x_k') - F(x_k-)| > l_0[\mu(e_0) - \epsilon_0] + l_1[\mu(e_1) - \epsilon_0 - \epsilon_1] + \dots \\ + l_{n-1}[\mu(e_n) - \epsilon_0 - \epsilon_1 - \dots - \epsilon_n].\end{aligned}$$

Then

$$\sum |F(x_k'+) - F(x_k-)| > \sum l_{i-1}\mu(e_i) - \epsilon$$

where ϵ is arbitrary, since $\epsilon_0, \dots, \epsilon_n$ are arbitrary independently of $l_0, \dots, l_{n-1}$ and independently of each other. Now x_k' may not be a point of $\mathfrak{E}'$. However, each of the intervals $[x_k, x_k']$ contains at least one point $\xi_k' \in \mathfrak{E}'$, which might be x_k, and it again follows from (5), that if ξ_k' is the first point of $\mathfrak{E}'$ to the left of x_k', then

$$\begin{aligned}(7) \qquad \sum |F(\xi_k'+) - F(x_k-)| &> \sum l_{i-1}\mu(e_i) - \epsilon. \\ &> 2M - \epsilon.\end{aligned}$$

But because $x_k, \xi_k' \in \mathfrak{E}'$, and because F is $AC - \omega$ over $\mathfrak{E}'$ it follows from (1) that the left side of (7) is not greater than M. We thus have a contradiction and it follows that $\sum l_{i-1}\mu(e_i)$ converges and consequently that f is Lebesgue–Stieltjes integrable over $\mathfrak{E}'$.

We now take a subdivision (l_{i-1}, l_i) of $(-\infty, \infty)$ which is such that if $e_i = E\{x : x \in \mathfrak{E}', l_{i-1} < f \leqslant l_i\}$ and n is sufficiently great,

$$\sum_{-n}^{n} l_{i-1}\mu e_i, \quad \sum_{-n}^{n} l_i\mu(e_i), \quad \int_{\mathfrak{E}'} f d\omega$$

do not differ by more than ϵ. For each point $x \in e_i$ there is a sequence of intervals $[x, x + h]$, $x + h$ points of continuity of ω, $h > 0$, $h \to 0$, and $\delta_x > 0$ so that

$$(8) \qquad \frac{F(x + h) - F(x-)}{\omega(x + h) - \omega(x-)} > l_{i-1}$$

$0 < h < \delta_x$.

Let (a_i, b_i) be the intervals complementary to $\mathfrak{E}'$ on $[l', m']$. Then because $a_i, b_i \in \mathfrak{E}'$ it follows from (2) that $\sum|F(b_i-) - F(a_i+)|$ converges. There is then an integer N such that

$$\sum_{n+1}^{\infty} \left| F(b_i-) - F(a_i+) \right| < \epsilon, \qquad n > N. \tag{9}$$

For n fixed, $n > N$, let $[r_m, s_m]$ be the closed intervals complementary to the set $(a_1, b_1), \ldots, (a_n, b_n)$. Then $\bigcup[r_m, s_m] \supset \mathfrak{E}'$. The method used to show that $|f|$ is Lebesgue–Stieltjes integrable with respect to ω over $\mathfrak{E}'$ can be used in conjunction with (8) to determine a set of mutually exclusive intervals corresponding to $[x_k, \xi_k']$ in (7), each interval on an interval of the set $[r_m, s_m]$ and with, if $x_k = \xi_k$, $\xi_k, \xi_k' \in \mathfrak{E}'$,

$$\begin{aligned} \sum F(\xi_{k'}'+) - F(\xi_k-) &> \sum l_{i-1}\mu(e_i) - \epsilon \\ &> \int_{\mathfrak{E}'} f d\omega - 2\epsilon \end{aligned} \tag{10}$$

and such that

$$\mu\{\bigcup \mathfrak{E}' \cap [\xi_j, \xi_{j'}']\} > \mu(\mathfrak{E}') - \epsilon', \tag{11}$$

where ϵ' is sufficiently small to insure that if $[\eta_i, \eta_i']$ is a set of non-overlapping intervals with $\eta_i, \eta_i' \in \mathfrak{E}'$ and $\mu[\bigcup\mathfrak{E}' \cap[\eta_i, \eta_i']]<\epsilon'$ then

$$\sum|F(\eta_i'+) - F(\eta_i-)| < \epsilon. \tag{12}$$

If in (10) ξ_k is for some m also a point s_m and a point of discontinuity of ω then $\xi_k' = \xi_k$.

Now consider the open intervals $(\bar{\xi}_k, \bar{\xi}_k')$ which are complementary to the set $[\xi_j, \xi_j']$ on the set $[r_m, s_m]$. Then from (11) $\mu[\bigcup\mathfrak{E}' \cap (\bar{\xi}_j, \bar{\xi}_j')] < \epsilon'$. If an end-point $\bar{\xi}_k$ of such an interval is not a limit point on the right of points of $\mathfrak{E}'$ then an interval $(\bar{\xi}_k, \eta_k)$, $\eta_k \in \mathfrak{E}'$, of the set (a_i, b_i), $i > n$, can be dropped from the left end of $(\bar{\xi}_k, \bar{\xi}_k')$ where $\mathfrak{E}' \cap (\bar{\xi}_k, \eta_k) = 0$. If $\bar{\xi}_k$ is a limit point on the right of points of $\mathfrak{E}'$, there is $\eta_k \in \mathfrak{E}'$, $\bar{\xi}_k < \eta_k < \bar{\xi}_k'$, with $|F(\eta_k-) - F(\bar{\xi}_k+)| < \epsilon_k$ when ϵ_k is arbitrary.

If $\bar{\xi}_k'$ is not a limit point on the left of points of $\mathfrak{E}'$ and interval $(\eta_k', \bar{\xi}_k')$ of the set (a_i, b_i), $i > n$, can be dropped from the right end of (η_k, ξ_k'), $\eta_k' \in \mathfrak{E}'$, $\mathfrak{E}' \cap (\eta_k, \bar{\xi}_k') = 0$. If $\bar{\xi}_k'$ is a limit point on the left of points of $\mathfrak{E}'$ there is $\eta_k' \in' \mathfrak{E}$, $\eta_k < \eta_k'$, $|F(\bar{\xi}_k'-) - F(\eta_k'+)| < \epsilon_k$. Denote the open intervals (ξ_k, η_k), (η_k', ξ_k') by (ν_p, ν_p'). Then, since ϵ_k is arbitrary,

$$\text{(13)} \qquad \sum|F(\nu_p'-) - F(\nu_p+)| < 2\sum\epsilon_k < \epsilon.$$

Furthermore, it follows from (11) and (12) that

$$\text{(14)} \qquad \sum|F(\eta_k'+) - F(\eta_k-)| < \epsilon.$$

We now have the interval $[l', m']$ covered by the mutually exclusive set of closed intervals $\bigcup[\xi_j, \xi_j']$ and $\bigcup[\eta_k, \eta_k']$ some of which may be points, and the open intervals $\bigcup(\nu_p, \nu_p')$ and $(a_1, v_1) \cup \ldots \cup (a_n, b_n)$ complementary to this mutually exclusive set of closed intervals. We then have

$$F(m') - F(l') = \sum \{F(\xi_j'+) - F(\xi_j-)\} + \sum_1^n \{F(b_i-)-F(a_i+)\}$$
$$+ \sum \{F(\eta_k'+) - F(\eta_k-)\} + \sum \{F(\nu_p'-) + F(\nu_p+)\}.$$

It then follows from (9), (10), (13), (14) that

$$F(m') - F(l') > \int_{\mathfrak{E}'} f d\omega + \sum \{F(b_i-) - F(a_i+)\} - 4\epsilon.$$

Then, since ϵ is arbitrary, it follows that

$$F(m') - F(l') \geqslant \int_{\mathfrak{E}'} f d\omega + \sum \{F(b_i-) - F(a_i+)\}.$$

The foregoing arguments can be used, in conjunction with the sets e_i where

$$e_i = E\{x : x \in \mathfrak{E}', l_{i-1} \leqslant f < l_i\}$$

to show that

$$F(m') - F(l') \leqslant \int_{\mathfrak{E}'} f d\omega + \sum \{F(b_i-) - F(a_i+)\}.$$

It then follows that

$$F(m') - F(l') = \int_{\mathfrak{E}'} f d\omega + \sum \{F(b_i-) - F(a_i+)\},$$

and because the points l', m' can be taken arbitrarily near l, m respectively and because F is $AC - \omega$ over $\mathfrak{E}$ it follows that

$$(15)\qquad F(m-) - F(l+) = \int_{\mathfrak{E}} f d\omega + \sum \{F(b_i-) - F(a_i+)\},$$

where $\bigcup (a_i, b_i) = \tilde{\mathfrak{E}} \cap (l, m)$.

We now return to the interval $[a', b']$. This interval is covered by $\bigcup\{[a', b'] \cap E_i\} = \bigcup \mathfrak{E}_i$ where each $\mathfrak{E}_i$ is closed and F is $AC - \omega$ over each $\mathfrak{E}_i$. There is then an interval (l, m) and an integer n such that $[a', b'] \cap (l, m) = \mathfrak{E}_n = (l, m)$. The function F is then $AC - \omega$ on (l, m) and

$$F(m-) - F(m+) = \int_{(l,m)} f d\omega.$$

Thus there are points x on $[a', b']$ which are in intervals (l, m) for which

$$(16)\qquad F(m-) - F(l+) = \int_{(l,m)} f d\omega.$$

Let G_1 be the set for which this is not true. Then G_1 is closed. Let (a_{1i}, b_{1i}) be the intervals complementary to G_1 on $[a', b']$. Then, if $a_{1i} < l' < m' < b_{1i}$, relation (16) holds for the interval (l', m'), l', m' points of continuity of F and

$$(17)\qquad F(b_{1i}-) - F(a_{1i}) = \lim_{l' \to a_{1i}, m' \to b_{1i}} \{F(m') - F(l')\}.$$

Now $G_1 = \bigcup G_1 \cap \mathfrak{E}_i = \bigcup g_i$ where each g_i is closed and F is $AC - \omega$ over each g_i. There is then an interval (l, m) and an integer n such that $G_1 \cap (l, m) = g_n \cap (l, m) = e$ where F is $AC - \omega$ over e. Reasoning such as that which led to (15) now gives

$$(18)\qquad F(m-) - F(l+) = \int_e f d\omega + \sum_{(l,m)} \{F(b_{1i}-) - F(a_{1i})\}.$$

Thus, there are points $x \in G_1$ which are in intervals (l, m) for which (18) holds. Let $G_2 \subset G_1$ be the set for which no such interval (l, m) exists. Then G_2 is closed. Let (a_{2i}, b_{2i}) be the intervals complementary to G_2 on $[a', b']$. If (l', m') is an interval for which $a_{2i} < l' < m' < b_{2i}$ with l', m' points of continuity of F and if $e = G_1 \cap (l', m')$ then, as in (18),

$$F(m') - F(l') = \int_e f d\omega + \sum_{(l',m')} \{F(b_{1i}-) - F(a_{1i}+)\}$$

and

$$F(b_{2i}-) - F(a_{2i}+) = \lim_{l' \to a_{2i},\, m' \to b_{2i}} \{F(m') - F(l')\}.$$

This process can be continued by transfinite induction in the usual way to arrive in at most a denumerable set of operation at $F(b') - F(a')$. Then

$$F(b-) - F(a+) = \lim_{a' \to a,\, b' \to b} F(b') - F(a').$$

This completes the proof of Theorem 3.4.

REFERENCES

1. N. Lusin, Sur les propriétés de l'intégrale de M. Denjoy. C.R. Acad. Sci., *155* (1912), 1475–8.
2. A. J. Ward, The Perron-Stieltjes integral, Math. Zeit., *41* (1936), 578–604.
3. R. Henstock, A new descriptive definition of the Ward integral, J. Lond. Math. Soc., *35* (1960), 43–8; N-Variation and N-variational integrals of set function, Proc. Lond. Math. Soc., *36* (1961), 109–32.
4. M. E. Munroe, Introduction to measure and integration, New York (1953).
5. R. L. Jeffery, Theory of functions of a real variable, 2nd ed., Toronto (1953).
6. E. C. Titchmarsh, Theory of functions, 2nd ed., Oxford (1939).
7. R. L. Jeffery, Generalized integrals with respect to functions of bounded variation, Can. J. Math., *10* (1958), 617–26.

SYMPOSIUM ON THE TRAINING IN MATHEMATICS GIVEN AT THE UNIVERSITY TO FUTURE HIGH SCHOOL TEACHERS

This symposium was held at the University of Montreal at 2:30 P.M., Friday, September 8, 1961. Copies of a notice (reproduced below) were sent to the Department of Physics and Chemistry in Canadian universities, and to each member of the Canadian Mathematical Congress.

NOTICE

THE TRAINING IN MATHEMATICS GIVEN AT THE UNIVERSITY TO HIGH SCHOOL TEACHERS

It is now recognized that the provision of competent high school teachers is one of the most serious problems in our educational structure.

One aspect of this problem lies almost entirely under the control of university professors, namely the *academic* training given to future high school teachers.

There is a growing feeling that the program of study at the university which is best for a future college professor is not the best for a future high school teacher and that the academic needs of the high school teachers have been neglected.

It is well known that the mathematics taught at the high school leads to its own particular problems in higher mathematics.

Should not the university offer courses on "High School Mathematics from a University Point of View" and give our high school teachers competence in their own area of activity?

And should not the mathematics and the science departments at the university cooperate more closely and offer a joint planned program of study for future "high school teachers of mathematics and science?"

Should the university offer to high school teachers of some experience who wish to improve their competence, courses designed for their needs?

The Canadian Mathematical Congress will focus attention on these questions by means of a panel discussion on the topic which is the title of this notice. The discussion will be held at Montreal on Friday, September 8, 1961, 2:30–5:00 P.M. It is expected that each of three or four key speakers will take 20 minutes to answer questions previously proposed to him by the chairman of the panel. Then there will be opinions and questions from the audience.

You and the members of your department are invited most cordially to participate in this discussion. More complete details should be available in a few weeks. In the meantime I would be glad to hear if there is any special activity going on in your department that should be reported to this panel discussion or if you have opinions you would like to present concerning the raising of the competence of high school teachers.

I. HALPERIN,
Chairman, Canadian Mathematical Congress Panel Discussion

Queen's University, Kingston
June 26, 1961

AFFICHE

FORMATION MATHÉMATIQUE DONNÉE À L'UNIVERSITÉ AUX MAÎTRES D'ÉCOLE SECONDAIRE

Il est maintenant reconnu que le recrutement de maîtres compétents pour nos écoles secondaires est l'un des problèmes les plus sérieux de notre système d'éducation.

Il y a un aspect de ce problème qui dépend presque entièrement des professeurs d'université, c'est celui de la formation académique des futurs maîtres d'écoles secondaires.

On est de plus en plus persuadé que le programme d'études universitaires le mieux adapté pour un futur professeur d'université ne l'est pas pour un futur maître d'école secondaire et que les besoins académiques de ces derniers ont été négligés.

Il est bien connu que les mathématiques enseignés dans les écoles secondaires posent des problèmes de mathématiques supérieures qui leur sont propres.

L'université ne devrait-elle pas offrir des cours sur le sujet « Mathématiques de niveau secondaire vue sous un aspect universitaire » pour donner à nos maîtres la compétence voulue dans leur propre sphère d'activite ?

D'autre part, les départements de sciences et de mathématiques de nos universités ne devraient-ils pas coopérer plus étroitement pour offrir un programme d'études conjoint pour les futurs « maîtres d'écoles secondaires de mathématiques et de sciences » ?

Et pour les maîtres d'écoles secondaires ayant une certaine expérience et désirant améliorer leur compétence académique, l'université ne devrait-elle pas offrir des cours appropriés à leur besoin ?

La Société Mathématique Canadienne veut attirer l'attention de chacun sur les questions par un « panel » sur le sujet indiqué par le titre de cet avis. La discussion aura lieu à Montréal le vendredi 8 septembre 1961, de 2:30–5:30 P.M. Il est prévu que trois ou quatre experts consacreront 20 minutes à répondre aux questions qui leur auront été proposées antérieurement par le président de la réunion. Ensuite l'auditoire pourra à son tour poser des questions ou exprimer leurs opinions.

Vous êtes très cordialement invités, ainsi que les membres de votre département, à participer à cette discussion. Des renseignements plus complets seront disponibles dans quelques semaines. D'ici là, je serais heureux de savoir s'il existe dans votre département des activités spéciales pouvant être mentionnées lors de cette discussion, ou si vous avez des opinions que vous voudriez présenter au sujet de l'amélioration de la compétence des maîtres d'écoles secondaires.

I. HALPERIN,
Président du panel
organisé par la Société Mathématique Canadienne

Queen's University, Kingston
le 7 juillet, 1961

Professor George Pólya was invited to address the Panel Discussion of the Canadian Mathematical Congress.

He replied that he was extremely interested but to his deep regret had to sail for Europe on August 21. However, at our request, he kindly sent a message, reproduced below:

To the Panel Discussion of the Canadian Mathematical Congress
Friday, September 8, 1961:

I wish to congratulate the Canadian Mathematical Congress for initiating the panel discussion to which these lines are addressed.

In fact, there is virtually unanimous agreement that the high schools should teach more mathematics and should teach it better. Yet it is much less clear how this goal should be attained.

The various groups now working on the revision of the curriculum deserve the thanks of the public. Yet two points should be clear: First, the proposed programs are experimental; some of their proposals may be eventually adopted, several will be rejected. To regard any of their proposals as final at this time, before further discussion and experimentation, to accept any one as a dogma, would be grievous error. Second, no change in the curriculum, however desirable it may be in itself, can cure all the ills: the real bottleneck is, in my opinion, the insufficient training of the teachers. Hence my best wishes to your panel which has the important task to discuss remedies for the sorest point of the high school mathematics instruction.

I wish to submit two points to your consideration to which I attached the greatest weight in the last five or six years in which my teaching activity was almost exclusively devoted to high school teachers. Each high school teacher of mathematics should have, I think, opportunity in his training for two things:

(1) Some practice in solving problems which need only knowledge on the high school level, but some degree of independence and originality. In fact, how should the teacher recognize or direct the creativity of his students if he himself had never an opportunity to do something approaching creative work?

(2) To acquire some knowledge of physical science, at least some serious understanding of elementary mechanics and optics. In fact, separated from science, mathematics loses one of its most important sources of interest and inspiration, not to mention usefulness.

I wish to mention that, in my instruction, I tried to take care of these two points which I consider as crucial by two sources. They are: (1) Seminar in Problem Solving, and (2) Mathematical Methods in Science.

My very best wishes for the success of your panel,

GEORGE PÓLYA

THE PANEL DISCUSSION

Chairman: I. HALPERIN, *Queen's University*

The speakers, Messrs. Bishop, Stanton, and Betts, were to discuss the following questions, posed by the Chairman of the Panel.

MR. A. W. BISHOP, *Ontario Department of Education*

Question 1: In a general way are the departments of education satisfied or dissatisfied with the curriculum now offered at the university to future high school teachers of mathematics and science?

Question 2: What opportunities are now available at the secondary school level to men and women trained in mathematics; as teachers, inspectors, textbook writers, staff members of departments of education? And to what extent is advancement likely to be based on ability rather than seniority?

PROFESSOR RALPH STANTON, *Mathematics Department, University of Waterloo*

Question 1: What action by Canadian universities would best assist in raising the competence of high school teachers of mathematics and science who are already teaching in the high schools?

Question 2: Should the university offer undergraduate courses of the type "high school mathematics from a university point of view" and if so, what material should be included in such courses?

PROFESSOR DONALD BETTS, *Physics Department, University of Alberta*

Question 1: Is it practical for the mathematics and science departments at the university to offer a joint programme of study for future high school teachers? What would be the advantages of such a programme, and what difficulties would have to be overcome?

Question 2: Is there at any of our Canadian universities any offering in this direction?

The Chairman opened the Panel Discussion by welcoming all present and called attention to the letter from Professor Pólya, emphasizing Professor Pólya's statement: "The real bottleneck is, in my opinion, the insufficient training of the teachers."

PROFESSOR HALPERIN

The invited speakers whom I have the pleasure to present to you are Mr. Bishop, who is an inspector in the area of mathematics with the Department of Education of Ontario, Professor Stanton of the Mathematics Department of the University of Waterloo, and Professor Betts of the Physics Department of the University of Alberta. Each of these speakers has kindly agreed to discuss specific questions which have been posed, and you all have copies of the questions.

Each speaker is giving his own opinion, and each one is responsible, of course, only for what he himself has to say.

After our speakers have discussed the questions, Professor Pounder, formerly of the Mathematics Department of the University of Toronto, now at York University, will open the discussion from the floor.

After that you are all invited to raise questions or to offer comments (critical or otherwise) or to speak your minds on whatever aspects of the problem you wish to discuss.

I would like to make one point. In the problem of how to improve our high school education there are many important topics which we *could* discuss: but today we propose to discuss only one of them. The topic that we are gathered to discuss is quite precise. It is a topic for which we bear almost the sole responsibility and we cannot brush off the responsibility or blame some other person or agencies for this aspect of the problem. This is the question of what we in the university do, to and for the students who come to the university and who end up teaching mathematics in the secondary schools. The training which they receive depends of course on their own abilities and their own willingness to work. But given these conditions, and they are certainly given to us, the question of whether we are doing the best that we could for them is the one that we are discussing now.

I will now call on our first speaker, Mr. Bishop.

MR. BISHOP

It is a very great honour to be asked to participate in this afternoon's programme.

This is just one more indication of the marked increase in the efforts by university personnel to understand and help solve the problems associated with the teaching of our high school students.

In our province, the Ontario Mathematics Commission is an outstanding example of the benefits to be gained from the three levels of education working together. The Commission was an outgrowth of an earlier committee on mathematics set up by the Ontario Teachers' Federation and broadened to make a special study of the high school mathematics curriculum. Set up first in

1959 as the Mathematics Commission of the Ontario Teachers' Federation and later reconstituted as the Ontario Mathematics Commission with representatives from the teaching federations, the Department of Education, the Ontario College of Education, and all the Ontario universities, it has made tremendous contributions in the field of secondary school mathematics. Although the members are not always in harmony, they have as a common purpose the improvement of our mathematics curriculum and the provision of teachers with the skills and attitudes necessary for developing our mathematically-minded boys and girls.

Before expressing an opinion on the questions posed to me I would like to say that we in the Department of Education who have responsibility for mathematics are particularly grateful to the Canadian Mathematical Congress for the encouragement that has been given to the better mathematical students who write our problems examination. For many years we have had a special grade 13 problems paper for students applying for scholarships in some of our universities. In each of the last four years the Congress has awarded a district prize of $50 for the highest problems mark in each of the 19 districts into which the province is divided for this purpose, and an additional $50 to each of the four or five highest marks on a province-wide basis. 698 students, which represents about 5 per cent of those taking grade 13 mathematics this year, wrote the examination. This competition has beeen one of the reasons for the encouraging increase in the number of problems classes and mathematics clubs in our schools in recent years. We thank you most sincerely for what you have done and we hope that your income continues to permit the support of this project.

I think now it would be useful if I provided those from other provinces with the procedure involved in becoming a teacher of mathematics in the secondary schools of Ontario.

The prospective teacher must first acquire a university degree, all courses being subject matter (the general Arts or Science degree is usually obtained in three years after grade 13 whereas the honour degree course requires four years), then attend the Ontario College of Education for one year or for two or three summer sessions to receive a teaching certificate valid in the secondary schools. On successfully completing the period of teacher training, the general-course graduate receives a high school assistant's certificate, type B, whereas the honour-degree person receives the high school assistant's certificate, type A, both valid for five years. Following two years of successful teaching, the former receives the permanent high school assistant's certificate and the latter the high school specialist's certificate.

Heads of our mathematics departments must be specialists in either mathematics or mathematics and physics, but there is no minimum requirement for completion of university mathematics courses to teach mathematics in our high schools. The holder of a high school assistant's certificate is legally qualified to teach any academic subject offered in our high schools. In actual practice, of course, the young teacher is assigned to the subject for which he is best fitted and in which he has spent some time of concentration at university.

To encourage up-grading we issue a teaching certificate endorsed in mathematics for 15 university mathematics credits (5 courses) in the degree work, and we issue a specialist certificate for 24 credits with 66 per cent standing, even though the honour degree course was not taken.

Since nearly all general arts or science students must pick a subject of concentration at university and since this accumulates 15 credits in one subject, it would seem that we might very well introduce a minimum qualification for the teaching of mathematics in our high schools—and that would be the teaching certificate endorsed in mathematics. Letters of permission for the teaching of mathematics could be issued to those not fully qualified in that subject, but the number of such letters would decrease year by year as the new basic certificate became established. For example, many of our physical education teachers would, in this case, require a letter of permission for mathematics teaching. They usually spend about one half of the day in teaching an academic subject in the class-room and by far the most popular is junior mathematics. In general, they are doing good work in mathematics because they are respected figures in the school, maintain good discipline, and demand regular work. Far too often, however, they have had no university mathematics and lack the skill and the imagination and over-all view of the subject which is so important in launching the eager young student into a voyage of discovery. There is a very good chance that the student entering your first-year mathematics classes received his start in algebra from a physical education specialist. I would urge then, that you do what you can to make it possible for students in the physical education course in the Ontario universities to take a course or two of the academic subject they are most likely to teach.

In answer to the question: "In a general way are the departments of education satisfied or dissatisfied with the curriculum now offered at the university to future high school teachers of mathematics and science?" I think I can say that our Department of Education is generally well pleased with the system of training mathematics teachers for the secondary schools. It is right that our future research people, astronomers, actuaries, statisticians, university teachers, and future high school specialists in mathematics and science should take classes side by side until the need for their own particular specialization arises.

We hope that you realize that the future high school mathematics teachers form a most important group in your classes because they are the ones who in the coming years will give the proper start and encouragement to an ever-increasing number of mathematicians and scientists. Thus, we hope that you will examine the content of every course you offer, with regard to the suitability for this group.

We want the honour degree person coming into our schools to be a scholar and a leader in his subject. But coupled with the importance of maintaining scholarship is the great need for specialists and we must have them in greater quantities.

The student coming out of grade 13 with first class or high second class honour standing in mathematics and science has every right to believe that he can succeed in an honour course leading to a teaching career. Yet many are failing. Does that mean that our standards are too low in grade 13 and we have given him the wrong picture of his ability in these subjects—or does it mean that the standards he is called upon to meet in university are unnecessarily high—or does it mean that we in the high schools have not developed in him the right attitude for independent thinking and research and a willingness to accept the responsibility to master a problem without supervision by a teacher?

We do need many more mathematics specialists in our schools. Few schools now can afford the luxury of assigning the specialist to grade 9 teaching—but that is where he is needed most. We need about 200 more to have an average of 1 per school.

We in the secondary schools must send you more of our capable people for your teaching stream in honour mathematics and physics. What can you do to send them back to us, trained in mathematics, in ever-increasing numbers?

Now, turning to the second question: "What opportunities are now available at the secondary school level to men and women trained in mathematics; as teachers, inspectors, textbook writers, staff members of departments of education? And to what extent is advancement likely to be based on ability rather than seniority?" I would point out that more than one-half of our schools do not have a mathematics specialist on the staff!

At the end of June we had in Ontario 426 provincial secondary schools ranging in size from Central Secondary School in Hamilton with 2268 pupils to Pelee Island Continuation School with 17 pupils. These schools were staffed by 11,500 full-time teachers. Nearly 2000 new teachers were required this fall—1100 to fill new class-rooms and about 900 to take the place of teachers who have retired or have withdrawn from the profession. These have been obtained and most (probably 85 per cent) have at least the minimum qualifications. Some class-rooms, however, are occupied by teachers without basic qualifications who have been issued letters of permission to teach.

The number of new teachers required will continue to increase year by year, and a pressing demand for fully qualified people will remain with us.

The person with a Type A certificate, the specialist, is of course a prized possession and greatly sought after. School boards most certainly recognize their worth and strive to get their share from the limited supply available.

Last year we had in our schools about 200 teachers endorsed in mathematics (5 university mathematics courses) and about 400 with the specialist certificate in mathematics or in mathematics and physics. This means that from 25 to 30 per cent of our mathematics teachers are specialists and another 12 to 15 per cent have at least

5 university mathematics courses. This picture looks quite bright but in reality the distribution through the province is uneven. 217 of the 426 schools do not have a mathematics specialist on the staff. Thus the opportunities for mathematics specialists are unlimited.

Our salaries are attractive, I think. The teacher with a general degree and a high school assistant's certificate starts at $4500 and will reach a maximum of $7800; with an endorsed certificate, from $4800 to $8200; with a specialist's certificate, from $5300 to $9100; with an honour degree and a specialist's certificate, from $6500 to $9500. The head of the department reaches a maximum in the area of $10,000 to $10,400; the vice-principal, $11,000; the principal, $14,000 in the large schools.

In general, promotion is on the basis of ability. Boards with several schools tend to select the best persons available within their area to become heads and vice-principals. The largest systems use the same plan in appointing principals from the ranks of vice-principals. In general, however, a school board advertises for a principal in a paper with a provincial circulation and gets applications from teachers, heads, vice-principals, and other principals in various parts of the province. It is interesting to note that 54 of our Ontario secondary school principals (1/8 of the total) hold the specialist's certificate in mathematics or mathematics and physics.

The heads of mathematics and science departments usually occupy favoured positions in competitions for promotion. Those of us who know their strengths in the class-room, and have special concern for the subject, often view with regret these well-deserved advancements to administrative posts.

This is a very good year to ask "What are the opportunities available for men and women trained in mathematics to become inspectors and staff members of the Department of Education?" For supervision of secondary schools, the province is divided into 12 districts with a district inspector in charge of the schools in his area. Until this year all the subject inspectors have lived in Toronto and have travelled the province, spending a week or two in each of the districts visiting their own subject in the class-rooms and meeting with their teachers. There have been two of us in mathematics. The department is now in the process of expansion. This fall, a newly appointed group of subject-inspectors will set up headquarters in London and will supervise their subjects in the schools of southwestern Ontario. It is probable that similar groups for Eastern and Northern Ontario will be set up in the near future. Thus, the answer is: "The opportunities have never been better."

What are the opportunities for the writing of textbooks? The Department of Education lists three or four texts for each subject in each of the grades 9 to 12. Schools are expected to select their texts from this list. There are no restrictions on the choice of grade 13 texts. Schools having a particular desire to use a non-authorized text for a special course or for a particular group or for experimental purposes may request permission from the Superintendent of Secondary Education and usually this permission is quite readily obtained. The arrangements for the experimental texts—mathematics 9 and mathematics 10—to be used in some of our schools were worked out on this basis. At the present time the opportunities for people trained in mathematics to write text books are quite good in Ontario.

Some of you will recall that in Ottawa in April 1960 at the seminar on "New Thinking in School Mathematics," sponsored by the Canadian Teachers' Federation, one of the resolutions passed read as follows: "The Seminar supports the idea of grouping pupils on the basis of their ability and performance in mathematics and advocates that multiple track plans be set up in schools in which this is administratively feasible."

Recently our Minister of Education announced plans for embarking on a much more formal streaming programme than we have had.

At present 76 per cent of our secondary school pupils are in the general course (the course which prepares for university entrance), 13 per cent in the commercial course, 9 per cent in the technical course, and about 2 per cent in special terminal courses.

Plans are now under way to organize schools, which have the proper facilities, into three branches which are being designated: the Arts and Science branch, the Business and Commerce branch, and the Science, Technology, and Trades branch. In each of these three branches it is proposed to have a five-year stream with courses of such a nature that the common grade 13 examinations at the end will secure university admission. Also, in each of the three branches there will be a less specialized four-year programme which leads to a graduation diploma at the end of grade 12 but which does not prepare for grade 13 and the university. In each of the last two branches there may be a two-year terminal group for those who have just barely made the grade from elementary school and also a one-year terminal group for pupils, at least 15 years of age, who have been transferred to the secondary school because of age.

Mathematics will occupy a central position in the curriculum for each of the groups in each of the branches and new courses and new textbooks will be required.

The secondary school system of Ontario needs many young men and women trained in mathematics. For able people who can teach, advancement is sure and speedy.

PROFESSOR STANTON

The two questions which have been posed to me are excellent questions but very difficult to answer. I shall attempt to give answers, but they will have to be subjective and undoubtedly not the only answers possible. I suspect that if a number of us tried to answer these questions, the number of answers would be of the same order of magnitude as the number of persons attempting to answer.

The first question, "What action by Canadian universities would best assist in raising the competence of high school teachers of mathematics and science who are already teaching in the high schools?" is one for which various solutions have been suggested. I think that I should start by saying that there are a few teachers for whom one cannot do anything. They are fortunately a small minority, and I do not propose to mention them at all. They are the ones that are simply not qualified, and cannot be made qualified in any way.

The vast majority of teachers might possibly benefit from one of three approaches, only one of which I think is really feasible. It would be nice to think that it would be possible for some of the mathematics teachers in the high schools to be granted leave of absence for a sabbatical year so that they might take a Master's programme at a university. And this is certainly desirable, where feasible. But I think that this device will probably not assume first rank importance because it is very difficult for Boards, considering the shortage of teachers at the present time and in the conceivable future, to allow their senior and best qualified people to take sabbatical years in any appreciable numbers. Also, it would be difficult for other than the more recent graduates to approach such a sabbatical Master's programme. If one compares a graduate of an honours course of about 1931 with one of 1961, the graduate of 1931 will probably have taken only about the equivalent of two of the four years of a present honours course. Therefore, he is not really in a position to enter upon a sabbatical Master's programme.

A second device that has been suggested, and indeed is being used, and which likewise has value up to a point, is the idea of teachers getting together for seminars a few times a year or joining for Saturday morning lectures on various topics. This is being done in many centres of our province. Frankly, I think that this device, though undoubtedly helpful, is of limited usefulness. A survey, of the nature that can be given in such a seminar, tends to be much more valuable to those teachers who already know the subject and are simply getting a bird's-eye view or review in such a discussion. It is difficult to treat any topic in a single lecture of that nature, unless it is of an historical type. Indeed, such seminars may even have a negative value for many teachers.

I have heard an experienced secondary school teacher give a talk in a seminar of this nature in which he valiantly attempted to present some facts about the axiomatic basis of the number system; but he simply did not have the mathematical background to do it. You cannot get yourself to the stage of being able to present such material by a few weeks or a few months of cramming. It requires a lifetime of work and experience.

I recall one seminar I myself gave to a group of teachers in Toronto on the iterative procedure for finding square root. I am not sure I did a great deal of service by it, as I am afraid many of the teachers merely regarded it as a useful gimmick that they would like to rush back and try out in the class-room, rather than a feature of mathematics.

In this direction, I would even question the value of some of the television programmes, such as "Continental Classroom," which are being presented at the present time, in which a great deal of abstract axiomatic mathematics is supposed to flow over the television waves into our secondary school teachers.

Seminars and lectures like this do serve a purpose in stimulating interest, but I do not think that they can serve any fundamental purpose because the value of presenting material to teachers lies, I believe, in two things—in the personal contact between the professor and the teachers with whom he is working, and, much more so, in the associated problems which are being worked. A single lecture, skimming over a topic, cannot serve the same role as a full-fledged course, given in depth, with a large selection of suitably chosen problems into which both teacher and professor can sink their teeth.

My opinion of the two devices that I have mentioned is that, though they possess a certain value, they do not carry us along very far in helping out the teachers.

What is the best way? My own conviction, and it is purely a personal view (probably half of you will agree and half will disagree) is that summer programmes for teachers, allowing them to take full-fledged courses in a mathematical discipline, are the only real basis for substantial improvement in the mathematical lot of our present group of high school teachers. We need these courses to give both breadth and depth. If the courses are interesting and well chosen, I feel that they can serve a very real need, and that the teachers can derive much of benefit from them.

But they must be well chosen. Simply offering a summer programme on say, the Peano Postulates, and taking as a text Landau's book, *The Foundation of Analysis*, is unlikely to inspire a teacher with anything but utter boredom. Fortunately, we have no such programme that I know of; that is why I selected this example.

Programmes of a suitable nature will help teachers keep up with modern developments in mathematics. I hesitate to use the word modern, as it is so ambiguous at the moment. When I use it personally I refer to developments in the usage of, say, modern algebra or digital computers, modern statistical theory, and so forth, rather than a change in symbolism. I have lived a number of years without ever making use of the symbol "upside down A," and I personally think that we will get a lot further with our high school teachers if we stress mathematics and the proper phrasing of questions in English rather than in symbolism.

If I might digress for a moment to give an illustration, during the past summer one of my own staff taught a course in numerical analysis to a group of about thirty-two high school teachers. The selection of numerical analysis as the course was probably due to my not entirely unprejudiced viewpoint that it is an excellent course for teachers, not only to keep up with modern developments and see how things have changed because of the advent of electronic computers, but also because of the fact that in such a course they can get quite a bit of breadth. Many of the teachers entering the course were well prepared; many were not well prepared. Some of them were physical education specialists, but because of two facts, namely, that it is a concrete, useful, down-to-earth course and because there is a great deal of problem material that can be associated with it, those teachers did a phenomenal amount of work, and went through the course very successfully. I think that, in any programme for teachers, these two features, *concrete usefulness* and *plenty of problems*, are the most important features.

In conclusion to the first question then, let me say that my belief is that the best way of helping teachers is to provide summer courses, well taught by experts, not by junior people. If that is done, I think that we will find that the teachers in high schools are excellent people, and we will develop an attitude of mutual respect between the teachers and the professors which will lead to even more fruitful co-operation than exists at the present time.

I pass on to the second question posed to me. "Should the university offer undergraduate courses of the type 'High School Mathematics from a University Point of View'?" I cannot answer "No," because I feel there is some value in the suggestion. Yet I cannot answer "Yes," because I feel that such an answer would be very ambiguous and might lead us into discussions of the rather dubious "Doctor of Arts" degree and the equally dubious "Ph.D. in Mathematics Education" which are being proposed in certain southern quarters.

It is possible that individual courses of such a nature are of considerable value if they are very carefully selected and very carefully policed. My personal preference would be to favour, not so much a course of this nature, as a programme of this nature, in which the entire four years of the programme were given with such an aim in view. To that end, I will describe what I think is a reasonable method to achieve something of this nature.

The first two years of an honours programme should, in my opinion, be common to all honours students, whether they are going into research, high school teaching, industrial development, university teaching, or any other field. This provides a basic core for all mathematicians. The things that I personally would like to see in those two years are: in the first year, a basic course in calculus, together with a course in algebra and vector geometry. In the second year, I would like to see courses in advanced analysis (differential and integral calculus), a first course in modern algebra, a basic course in probability and statistics, and an option between courses in mechanics and actuarial science.

Because of personal predilections, I am opposed to compulsory courses in actuarial science, a subject for which I have a limited affection. I have found that there is little difficulty in optioning it with mechanics, since very few students who wish to take mechanics wish to take actuarial science, and vice versa.

I think that, from there on, in the higher years, one must try to achieve two things: a reasonable depth, but considerable breadth; for a person going into high school teaching, breadth, once he has the basic material, is more important than depth.

If one concentrates on what I call *useful mathematics*, by which I do not mean mathematical logic, then I think that one can do a great deal for persons entering a teaching career. For example, in the third year of a programme, I would feel that every teacher should have a course in the real and complex variable which would carry them up to, say, the Cauchy Integral Theorems. For a student going into high school teaching, I think that it would be far more valuable in his fourth year, rather than to continue to a second course on advanced work in the complex variable, to take a course in a different field, so that, while not being quite as specialized, he would have a broader spectrum of knowledge. I think that this would help him greatly both in his presentation and in his work with students.

If you will permit me to mention a personal illustration, we have found at Waterloo that it is very useful to introduce, in the third and fourth years, what we call a *teaching option* which has considerably more flexibility than some of the other options. In the third year, students selecting this option take their second course in modern algebra, a course in Euclidean and projective geometry, a course in numerical analysis, and a basic course in complex variables. They have one other mathematical option.

But in the fourth year we demand only one compulsory course. We demand that they take a course in the history of mathematics, which is not a list of names and dates but a course in which they study, not only the men who created mathematics, but also what they created. There is a considerable amount of mathematical content, including problem solving, in this course, and I think that the students learn quite a bit by seeing how different problems were solved at different eras in the development of mathematics. Their other final year courses, four in number, are all optional. We do not demand that they take a second course in complex variables, nor do we demand that they take the third course in modern algebra. In fact, we feel that for teachers going into high school work it is far more important, rather than taking one of those very advanced courses, to go back to the second year programme and pick either mechanics or actuarial science, whichever was not selected as an option at that time.

I think that, in a teaching option of this nature, the courses from which the students derive most benefit are courses like that in the history of mathematics which improves not only their mathematics, but, by essay writing, their English; the course in complex variables which gives them a taste of higher analysis; the two courses in modern algebra, where we try to stress not a sterile symbolism but the solving of a great many problems; the course in numerical analysis where they gain experience in various branches of mathematics, and where they can be very concrete and very modern at the same time; and, finally, the course in geometry, where they find a vast number of concrete and interesting problems. I cannot help mentioning at this time, in case anyone has not read Professor Coxeter's new book, *Introduction to Geometry*, that it is a long time since I have read a book where I found such a fascinating presentation of mathematics. I think that each of our students going into the teaching profession should own a copy of that book; it is a magnificent production.

In conclusion to the second question, I think I would say that our best approach in the university is to offer for the teachers a sound programme with good breadth and sufficient depth, and I think that an able professor who knows his field and is enthusiastic about it can use almost any branch of mathematics to illustrate good sound mathematical thinking, provided he sticks to the two ideas of *problem solving* and *concrete application.*

PROFESSOR BETTS

I am not sure why I, an undistinguished physicist, have been asked to participate on this panel when there are so many distinguished mathematicians here, but perhaps I can contribute from a somewhat different point of view. I hope you will not be displeased if I do not spend all my time on answering my two questions.

That the great majority of mathematics and science teachers in high school are not really competent in these subjects is I think well known in university circles, but is insufficiently appreciated by the general public. In my own experience, in all my years as a schoolboy, I considered then and I consider now, that I had only one mathematics teacher who was of any value to me. The rest of the time I would have been better off reading a good textbook on my own. I would gather from Mr. Bishop's remarks that the situation is much better in Ontario and I am very pleased that this is so.

I have been reading over the past few days a large number of university calendars, particularly the sections under Education Faculty or Education Department, and I would like to mention now, very briefly, how it seems to me that in practice teachers are trained.

In what I consider the best cases, the high school mathematics or science teachers have had honours Arts or Science training in mathematics or some science subject, necessarily followed by a year of education courses. I think across Canada it is impossible to teach in schools without having a year of education courses. Very often though, a person has proceeded straight to a Bachelor of Education degree, taking mathematics or science only as his education speciality. Worse, a person who is teaching mathematics or science may not even have had any mathematics or science while he was taking his Bachelor of Education, and he may not even have a Bachelor of Education degree because he may have taken only one year of education and then be out teaching. I would like to point out that those who teach physics or any other science are in general worse off than those who teach mathematics. For the purposes of categorization in educational faculties, the sciences are all lumped together in one category which may even include geography. Once the teachers specialize in science, which may mean, for example, that they have had only two university physics courses and perhaps one each of geography, geology, and chemistry, they are science specialists.

I am supposed to answer the question, "Is it practical for the mathematics and science departments at the university to offer a joint programme for future high school teachers?" First of all, one has to recognize that the departments of education control the certification of teachers and require education degrees for permanent certificates, and university faculties of education in general control the programmes which the teachers take. So, while mathematics and science departments may offer as much as they like, in general not much good is done unless there is collaboration between the education faculties and the mathematics and science departments. This is particularly true in universities where the students enter education faculties directly from high school rather than after taking bachelor degrees.

As far as the obstacles to any such programme are concerned, there are obstacles on all sides, some of them those which are due to the university people like ourselves and also some others that I would like to mention.

We cannot blame the education faculty altogether. The mathematics and science departments, as far as I know, have shown little or no interest in training high school teachers. If a prospective teacher wants to come along and take a course that they are offering, all right; he may sit there and pass it if he can, but no special interest is taken in him.

There is also the attitude of the students. The attitude prevalent is that a good student does not go into education. I remember a panel discussion at which an educationist (one of the few contacts we had with the educationists) suggested to one of our best students that he go into teaching after finishing his honours degree. The students looked at him as if he had holes in his head, and then everyone looked very awkward as if they were saying, "What an idea! This is a good student. We send our poor students there!" Now this attitude among the students is one for which we are to blame. We need to do something to counteract it because teaching is a very important occupation. We will not get good students from the high schools if we do not send some of our good honour students to education.

There seems to be little reward for those who have taken special courses. I know that in some provinces the salary which is paid depends only on the number of years of service and the number of years of education, regardless of the particular specialized competence of the teacher. There is only one way in which special competence can be rewarded: in the larger schools teachers can be made heads of departments, and in smaller ones, they can be made principals. Then they are not teachers any more!

Now I shall try to answer the specific questions that I have been asked. I do not believe that there should be a joint programme in mathematics and science. I believe that there should be a special programme in mathematics more or less as Professor Stanton has outlined, and there should be similar ones for *each* of the different sciences. I think that anyone who wants to teach high school physics has to know what physics is about. By the time that he has even an honours degree in physics, he has taken only one course in each of the important fields of physics, and can have hardly any idea what modern physics is about unless he has an honours degree in physics. He could take some special courses along the way. But I do not think a joint programme in science and mathematics would be suitable because students would not have enough courses in physics to teach physics properly to high school students.

As far as the second question is concerned, "Is there at any of our Canadian universities any offering in this direction?" (i.e., a joint programme of specialization for science and mathematics) there are majors and minors in education faculties. Very often there will be a major in mathematics which requires a student of education to take during his education course five courses in mathematics. Then he might minor in science, which means he might take as many as two courses in physics, or vice versa, but that is about all I have been able to discover. I do not think that is quite the same sort of thing as a joint programme of specialization.

Perhaps I might end on a constructive note. First of all I would like to see students receiving financial aid (comparable to the aid given to education students) in order to take Arts and Science degrees in a subject as long as they were looking forward to going into teaching. As it is, if students go directly into Education, they receive very generous financial assistance compared to the assistance given to those who go straight through the honours programme; hence students tend to go to the education faculty directly if they want to become teachers. Now I do not think Education is the best route for high school teachers.

I think prospective teachers should be able to look forward to much higher salaries than they have now. We have heard of what seems to be a generous salary scale in Ontario, but I do not think that even it is high enough. We get students who find that going into the Defence Research Board is more attractive financially than going into the teaching profession. I think the teaching profession is more important than the Defence Research Board.

I think we must try to cultivate in the public an attitude that the training is not sufficient now and must be better. The public, I think, has probably been responsible for the fact that people who teach manual training or household economics must have specialized certificates, and that there are some faculties of education that give about eight different specialist's certificates. However the certificates are all in subjects like manual training where the public can readily see whether a person has an adequate ability to teach his subject. Almost anyone can see if a chair is well made, but not everyone in the general public can tell whether a person has been taught properly in mathematics or physics or some other science; therefore there is not the same demand for really good teaching.

We can try to foster that demand. We can ask that permanent certificates be given only to those who have at least an honours bachelor degree in the subject they are going to teach in high school. We can set up in mathematics and physics departments courses like "elementary physics from an advanced standpoint" or "elementary mathematics from an advanced standpoint" and so on. We can appoint in our departments someone with particular responsibility for seeing that there is some liaison and some improvement in relations between the professional educators and the mathematicians or physicists as the case may be. We should encourage in the teaching profession a more responsible attitude. In Alberta so-called Speciality Councils have recently been set up in which the teachers get together to discuss their problems, hold seminars, and so on. It is hoped that the Speciality Councils will help develop the professional attitude and make the teachers themselves more aware of the fact that they need to have a much better education in their subject matter.

PROFESSOR POUNDER (*York University*)

Mr. Chairman. I am going to offer a few personal opinions about what our universities should do to improve the training of future high school teachers. Because there is a practical limitation on the number of courses that a university department can provide, it is to be assumed that the prospective high school teachers will be trained in courses that involve compromises between their interests and those of other groups.

It is my thesis that the situation for the majority of prospective teachers of mathematics would be much improved if the area covered at present were reduced by a quarter and the time thereby released to the student were devoted by him to the solving of problems appropriate to his studies. The subject-matter of our courses may be very good, but candour requires us to admit that some of our most advanced undergraduate courses are not adequately understood by a considerable fraction of those nominally studying them. Many students in this category would secure better academic qualifications than they now obtain if they achieved a reasonable mastery in the more modest programme suggested above.

I would be pleased to see much problem work required and a more complete control demanded. The problems should be custom-made to fit both the development of the subject and the capacity of the student. What heightens interest in the student is, on the one hand, the sense of having done something by himself, and, on the other hand, the evident interest and satisfaction of his teacher in what he has done. It is an important object, in the universities as in the schools, to arouse and maintain interest in a subject. It is probable that there is no way of attaining this object as successful as that of providing the student with appropriate problems and considering his efforts with him. Also, this consideration tends to keep the instructor's feet on the ground. I agree completely with Professor Stanton that the problems should appear to the student to be natural ones and worth working at. They should be within his powers but the solutions should be so far from obvious to him that his discovery of them will induce a feeling of pleasure.

It may be expected that when the student becomes a teacher he will try to have his pupils obtain a mastery of their studies in the way that he found effective for himself. His mastery in his own studies will give him a confidence in his own judgment very desirable in the present period of discussion and change and new programmes.

We may ask how it is that we try to have our students in mathematics take courses, in number and in depth, beyond the powers of many of them. The reason appears to me to be that we feel that we must keep up with the Joneses, that our level of treatment ought not to be lower than that attempted elsewhere. If, however, the result is, as I believe it to be, a considerable measure of failure, we ought to suppress our impulse for more advanced work and to prefer genuine accomplishment by the majority at a modest level to an exposure to ideas at a high level that relatively few have the capacity to grasp. By dealing with these few privately and by insisting on a greater mastery by the majority of subject-matter accessible to them we shall be attending to the former group and helping to provide a supply of teachers both more numerous and better trained.

PROFESSOR LEON HENKIN (*University of California, Berkeley*)

I should like to mention another way, somewhat related but rather different from those mentioned previously, in which it seems to me the university can play an important role in raising the competence of our mathematics teachers.

In thinking of this question, it is natural first of all to imagine how we might assist in terms of the university as a teaching institution; yet, when we stop to think of it, the university plays a much wider role in our society. In particular it does, I believe, in your Canadian provinces as in our American states, play an important role in moulding public opinion, drawing the attention of public opinion to important questions of education and scholarship.

It seems to me that there is no more important thing that the university can do in raising the competence of teachers than to grasp this opportunity for leadership in ameliorating the terrible conditions of work under which most schoolteachers labour. I refer to the number of contact hours, the number of hours spent each week in dealing directly with students. Let me explain.

I think we at the university, in so far as our own instructional programmes are concerned, have long recognized that a really adequate teacher must, of necessity, do many other things than merely teach in class-rooms. The really adequate teacher *must be fully engaged with his subject*, must have time to think about it—not merely in his period of training, but every day. He must continually try to see his subject in a new way, must try his best to reformulate it creatively to whatever extent his abilities may permit, and to read what others are doing in the field. When the teacher is engaged in this way he brings to his instructional task a measure of enthusiasm which is by far the greatest secret of the successful teacher. Even if a teacher's knowledge is limited, if he is excited about the subject and if he can get his students excited, the students will make great progress by themselves.

But when you have teachers who are burdened down by twenty-five hours each week of contact teaching, followed by reams and reams of clerical activities such as the grading of papers and the submission of records, followed by various extra-curricular tasks that are superimposed on them, the poor creatures simply have no vitality left with which to maintain a continuous probing of the intellectual aspects of their subject. It seems to me that despite the great shortage of teachers we must immediately make every effort to shorten teaching loads. By doing this we will not only improve the performance of our teachers but we will make the teaching profession more attractive, and we will begin to correct the shortage from which the profession now suffers so seriously.

PROFESSOR ROBINSON (*University of Toronto*)

I have been making some notes of what the various speakers have said, since I have been involved in this problem during this past year to a painful extent. The University of Toronto has been trying to do something about it—in spite of what Professor Betts says—in a series of six meetings with members of the mathematics and scientific departments of the Ontario College of Education. Mr. Bishop may have heard of these meetings; whether they will be productive of results or not remains to be seen. Another place where we have been trying to make some progress is in a joint committee with the Toronto School Board. Thus at least one university has been taking its responsibility seriously with regard to the shortage of teachers.

It is not a shortage of bodies, but a shortage of trained teachers and this is one of the curious things about the situation. On numerous occasions, the public has been told that there is no shortage of teachers! At least we have been able to correct this false impression, since the statistics to which Mr. Bishop referred indicate clearly that only about 25 per cent of the high school teachers of mathematics and science have had a proper training in their subjects.

One thing has not been mentioned with regard to mathematics. In the statistics of the Dominion Department of Labour, the demand for mathematically trained people extends throughout our whole society, not just in the teaching profession. What surprised me was that it exceeds that for chemists, physicists, engineers, or for any other groups with specialized training. The explanation is clearly the enormous increase in the use of electronic computers. Nearly every industry is installing these machines and the larger the industry the better the trained personnel which it requires. Now it is all very well to say that the universities should look on prospective teachers as a privileged group in our classes. But this simply does not satisfy the needs of the community at large. We have the responsibility of providing university teachers, people our industrialized society requires, and we have to train teachers for the high schools. Here is the nub of the problem: how can universities meet these three distinct demands? This is what our committees at the universities have been struggling with. How can we reconcile the courses we give to the requirements of the different groups?

There is another problem that has not been mentioned: it lies in the teachers' organizations themselves. Thirty years ago a man took a mathematics course and he could become an actuary in an insurance company, or he could become a teacher in a school or university, but the universities were smaller and the number entering university teaching was very much less. Now the picture is different. The opportunities for a graduate are as I have described and the teaching profession sees less qualified teachers coming into the schools. The well qualified teachers want to protect their status. We want to do the same in the universities; we want to protect the title professor against interpretations which might lower its importance. The high school specialist has the same attitude, and this factor works against a realistic approach to the problem. It is a difficult situation. One can see both sides of the question very clearly and one has to approach it with understanding. Nevertheless, the facts of the case are as they have been stated by nearly every speaker, that the proper training for a high school teacher is not that of the specialized courses that are necessary for the man who is to go on to university teaching.

We have been struggling to try to reach some sort of a solution in which teachers will recognize and accept these changes in our society. They should have the courses that Professor Stanton has described in order to give them a proper breadth of interest and understanding. We need the co-operation not only of the teachers but also of all departments in the universities if we are to make progress.

One other point has not been mentioned. It is so important that the College of Education asked me (I was chairman of the committee in question) to interview the principal of one of our large high schools, asking him this question: How would you compare the value in the schools of the specialist in, say, mathematics only, with the teacher who has had the appropriate training in more than one subject?

Now this is critical, particularly in these days when shortages are so acute. If you have a school, as Professor Bishop said, with no specialist in mathematics, who is going to teach the upper grades? If recourse cannot be had to the teacher of physics, it is going to be the physical health man who does the teaching. And in the same way, if the school happens to have a mathematics teacher who has the qualifications and no teacher of physics, who is going to teach the physics? I came away from that interview with a better understanding of the problems of the schools, but I must say that these problems are not as acute as they once were. In the Province of Ontario, the schools have been integrated. There are sectional high schools in the country and most schools are larger than they were even ten years ago, with several qualified teachers. I think you could say that a specialist in one of these schools will teach *two* subjects, whereas previously he would have had *three*. That is about the level of change, and in a big city school he may teach only *one*.

It is exceedingly important that all of us take these problems seriously. It is easy to criticize colleges of education, and I am the first to recognize their faults, but we are all seeking a solution to the same problem and the one thing that is necessary is for us to get together on it.

PROFESSOR DIVINSKY (*University of British Columbia*)

I have taught high school teachers at the University of Oregon for the last two summers. I have had some contact with them in Manitoba and now I have had some contact with them in British Columbia and I feel that their level of achievement is much lower than what has been described up to now. When I hear talk about second and *third* courses in modern algebra for prospective high school teachers, I simply do not understand at all.

I gave a first course in modern algebra to a very carefully selected National Science Foundation group of high school teachers; I think 60 of them were selected from 1600, from all over the United States, and even one from Japan. And at the end of this course, the *first* course in algebra, they got almost nothing out of it. This last summer I gave them a course in the theory of equations which they more or less understood. They had the feeling of accomplishment that was described and I think this course was much more successful. But I would like to speak up for saving our honours programmes in the universities, and I would disagree very strongly with Professor Stanton in regard to some of the courses that he described.

PROFESSOR WITTENBERG (*Laval University*)

I would like to quote an old German saying which in our context might go somewhat like this: "Sweep at your own door before you sweep at the door of the Faculty of Education." To put it bluntly, I think there is a very simple reason for the difficulties we have in inducing the people in the schools, and the future teachers, to take the teaching of mathematics seriously. It is that, despite all the appearances to the contrary, we ourselves, the university people, have failed in our responsibility to take the problems of teaching mathematics in schools seriously to any reasonable extent. And I shall explain precisely what I mean by that.

What does it mean, for us, to take a problem seriously? We know very well from our professional work: it doesn't mean to devote time, it doesn't mean to devote money—it means to devote *thought* and serious scholarly consideration to the problem. Professor Henkin referred a moment ago to the professional, whose own subject is always in his mind and who is constantly involved with it. I believe this also applies to the problems of mathematical education. We do take them seriously to the precise extent that we give them this kind of devotion. We do *not* take them seriously to the extent that we deny them this kind of devotion.

In fact, we did deny it to them to a very large extent in the past. And by so doing, we also denied it to the future teachers whom we train. How can we expect them to give serious thought to what, obviously, we do not deem worthy of our own serious effort?

This is not only a matter of professional training. It is a psychological matter as well. What we indirectly (and perhaps sometimes even directly) suggest to the future teacher, is that reflection about the teaching of mathematics in schools does not raise intellectual challenges of a magnitude and of an importance comparable in any way to those raised by mathematical problems proper. We tell them by our deeds, if not in so many words, that there is nothing really exciting about the problems of mathematical education, and that solving them is a kind of odd job to be left for odd moments, week-ends perhaps, or the summer when it is very hot and we are too tired for what we consider more important work. And the suggestion, of course, is self-fulfilling.

Let me be quite specific: you know that for years a great hullabaloo has been going on concerning the teaching of mathematics, that millions of dollars have been spent, countless courses given, publications written, and so on. And yet I shall draw your attention to a very simple fact: even today, the very first element of serious scholarly discussion of the issues involved is wanting—namely, a journal, or a part of a journal, that would be devoted to discussing the fundamental assumptions on which all this work rests. Take for instance the journal that ought to be the main organ in this respect: the *American Mathematical Monthly*. If you look through the "Mathematical Education Notes" in it, you will find that at least 90 per cent of the space is given to describing what people *do*, not what people *think*. I shall not proceed from here to the conclusion that, in fact, some of the people in question have not thought, but have just done, though sometimes this conclusion would not be entirely unwarranted. (The underlying principle might, with some reason, be called an extension of Parkinson's Law: *act now—think later*!)

Generally speaking, this results in a situation in which there is an astonishing disproportion between the amount of work done concerning the teaching of mathematics on the one hand, and the amount of serious and critical reflection concerning the assumptions underlying this work on the other.

For instance (I may be mistaken, and I would be glad to be corrected), I do not think that there has ever been any serious, responsible, scholarly public discussion of the so-called modern approach to the teaching of mathematics. Concerning this approach, there are a number of things to be said against it, and a number of things to be said in its favour; this is not the point I want to discuss now. My point is that questions such as these ought to be aired publicly and with intellectual responsibility if we ourselves want to claim to have given serious attention to the problems of mathematical education. There is not one among us who would think for one second of publishing a paper about algebra or an advanced textbook or anything of that kind without giving very serious, prolonged, and devoted thought to it. But today it seems perfectly normal that nothing of this type is necessary for work about the problems of teaching. This is, in my view, an intolerable situation—the more intolerable, since in fact these problems, if different, are no less challenging and complex than any to be encountered in the realm of pure mathematics.

To present the same remark in a slightly different, but substantially similar context: it is an axiom that not one of us would challenge, that no graduate department of mathematics could do anything useful, if it were not solidly rooted in research. But we shy away from drawing the same conclusion with respect to teaching matters: the conclusion that it is nonsensical to train future teachers of mathematics in a department where nobody has ever done any serious thinking whatever concerning the what, the how, the why, and the wherefore of the teaching of mathematics in schools. I think that this is a very important contributing factor to the mess in which we find ourselves.

Of course, nobody would deny that the "educational research" that has been done concerning the teaching of mathematics has often been extremely bad. But why was it so bad? Because those people who could have done it better have not seen fit to do it.

To quote a familiar example: you will often hear proponents of "modern" reforms state, as Professor Henkin did here, that these reforms are essential if the pupil is to "understand the modern world," or something to that effect. But if you ask them what exactly are the important aspects of the modern world that the pupil will understand if, and only if, he studies the suggested subjects, you will find that they never gave thought to this seemingly so natural, if not trivial, question; a question that would have been the very first to come up in any serious seminar about the teaching of modern mathematics.

Is it not a situation of intellectual irresponsibility when we keep ourselves so busy peddling reforms that we do not find the time to reflect whether these reforms are sound?

My conclusion would be: if we really are concerned with improving the teaching of mathematics in schools, our very first duty is to create a situation in which this concern will be translated into responsible and scholarly effort, performed under intellectual standards that are no less exacting than those accepted in mathematical fields. This means, on the one hand, creating, in several university departments of mathematics, centres of deep reflection about the aims and methods of the teaching of mathematics in schools. No proposals for reform that are not demonstrably rooted in such a reflection should attain a status of intellectual respectability. On the other hand, it means the creation of an international journal devoted to the teaching of mathematics, edited by responsible scholars of unquestioned stature, free of any entanglements with specific reform movements, and with the sole aim of promoting the broadest and most thorough inquiry into all aspects of this problem.

ADDED APRIL 1962

The recent statement by 65 mathematicians, "On the Mathematics Curriculum of the High School" in the March 1962 issues of the *American Mathematical Monthly* and the *Mathematics Teacher*, confirms and corroborates the above remarks. (Significantly enough, the latter journal put the statement in a column headed "A Column of Unofficial Comment"!) On the one hand, it shows that for years highly promoted and heavily subsidized reform movements worked in directions that were considered unsound by, among others, mathematicians of the distinction of Professor Courant and Professor Pólya, without this fact ever becoming apparent—as it would have become immediately had there been the scholarly discussion of the issues involved, referred to in the above remarks. (In fact, it seems that, while elaborate pains were taken to ask high school teachers in Kalamazoo what they thought of the experimental materials, no similar pains were taken to get an appraisal from men like those just named.) Unfortunately the statement also shows something else: by confining itself to the formulation of guidelines, by being unable, after so many years, to point approvingly to one single example of curriculum reform as a model and a standard to be followed, the statement makes drastically clear how little has really been and is being done.

The reluctance of many of those involved in the reform movements to get involved in critical discussions of their assumptions and aims, their timidity in thinking critically, and haste in *doing*, are sometimes quite remarkable, both in Europe and in America. If I may be permitted to mention a personal example: I published various very outspoken critical remarks challenging several of those assumptions in Germany, in England, and, by implication, in France and Belgium. Not one of these papers provoked a single reaction in the way of criticism, rebuttal, or even discussion.

A very thorough discussion of some aspects of the teaching of mathematics along lines closely related to those postulated in the above-mentioned statement, will be contained in my forthcoming book, *Bildung und Mathematik*, to be published by Klett in Stuttgart, Germany.

RESEARCH POLICY SEMINAR

Chairman: G. DE B. ROBINSON, *University of Toronto*

CHAIRMAN

Last night some of us were discussing methods of research and of stimulating research. Two of the replies of two notable members of this seminar were to the effect that you sat down and did it. But it is a little more than that because the matter of policy does enter into the picture and some will say that the organization of research, inside and outside our universities, is important. In the English-speaking world the tendency has been to have research done within the universities, though there are notable exceptions—the institutes of Princeton and Dublin. We all know that in many universities there are professors who do no undergraduate teaching, so that the picture is not clear. Nevertheless, outside the English-speaking world there is greater emphasis on segregation of research from undergraduate teaching. This is policy, and there is also the stimulation of research by example and by personal contact, which has nothing to do with policy. These are the two aspects of the problem which I hope you will consider.

R. L. JEFFERY

When the question came up for a panel discussion on ways and means of advancing research in mathematics, I was pleased to say something. It is all I have thought about and worked at for many years. But when I came to set down what I might say it was clear that working and thinking does not necessarily make one wise. I shall begin by pointing out that our summer vacation in Canada is longer than that in other countries. This has been exploited by the sciences, other than mathematics. Considering her late start and limited resources, it is remarkable how important a place Canada has made for herself in the natural sciences. I do not wish to leave this point without paying tribute to the men involved, most of whom I have known personally. For two generations they have worked in their laboratories through the long summer months without recompense or the thought of recompense, apart from the satisfaction of helping students in advancing knowledge. There have been no government handouts from which university staff members could benefit personally nor are there today; not in Canada. Only during the last twelve years have mathematicians taken advantage of the summers in any organized way through the biennial seminars, which many of you have attended, and the summer research institute. Fewer of you have served time at the institute which is a three-month stretch from mid-May until mid-August. A succinct description of the purpose of the institute is found in a report of the Conference on the Support of Mathematics of the Federal Government of the United States. One of its recommendations is: "It is desirable to establish summer research seminars largely without organized courses (following the lines of the annual summer Research Institute of the Canadian Mathematical Congress) in which mathematicians from institutions of various types can meet together with adequate support to work on research problems and discuss their ideas. In such a seminar, younger mathematicians would be able to devote their full energies to research, faculty members from isolated institutions could be re-vitalized by contact with enthusiastic research forces, and leaders in Mathematics could make contact with interested younger colleagues." I don't know how they know so much, but they pretty well hit the nail on the head by the description of what the summer research institute is.

In the summer of 1950, staff members in mathematics departments in Canadian universities were carrying loads of routine work that left little time, if any, for independent study. Salaries were such that most staff members felt, and justifiably, that they had to take paying jobs during the summer if they could find them. In only three of our universities was there anything approaching a library, although this is not a serious matter in mathematics. It was to alleviate this situation that the Institute with its modest fellowships was established, and it is generally agreed that it has had a good measure of success. This past summer there were 41 fellows, and a dozen or more visitors. Professor H. Nakano of Japan was a fellow, and Professor Nagy of Hungary and Professor Grätzer of the Mathematics Research Institute of the Hungarian Academy of Sciences were visitors. Professor Nagy gave a series of lectures which were well attended. In general, however, there are few formal lectures but small groups do form to talk about a particular topic in which the members are interested.

Much thought and discussion has been given to trying to formulate procedures that would increase the effectiveness of the Institute. But so far every proposal has carried the risk of destroying to some extent the values already achieved. I do not want to say too much about the Institute, but if it is to continue it must have increased accommodation, and increased grants for fellowships. To get these will require the serious moral support of the whole mathematical community. As I look back over the last 12 years, I realize that our problems have been eased considerably by the unanimous and generous encouragement of the officers and members of the Congress and I want to take this opportunity to express my deep gratitude to all. It is very important that you give the same generous support to those who will be carrying the burden in the future. I am sure we are all deeply grateful to the National Research Council as well for its generous support, and to those American foundations which have contributed so liberally.

With this said about our Summer Research Institute I now have a confession to make and an apology to some of my friends. For a long time now there has been agitation for a full-dress national institute in mathematics. Many of my friends who I am sure are wiser than I have supported it, while I have opposed it. A national institute is not without some merit *per se*. It would be an ivory tower with a glass dome in which a few gifted individuals could bask in perpetual intellectual sunshine. They and their doings would be taken by the world at large to represent the level of mathematical achievement in Canada, no matter how blighted the conditions in our universities should remain or become. More than any other Western nation, Canada is developing sectionally. For a long time to come all available resources for education, both federal and provincial, will be needed to further this development. Such has been the pattern to date. It has been generously and intelligently supported by the National Research Council, and now the Canada Council is lending a hand. Considering our late start, our limited resources, and the isolation of our universities one from the other, the extent to which a real university atmosphere prevails is remarkable.

President H. G. Thode of McMaster, in his convocation address at Acadia, pointed out that in Russia there were hundreds of research institutes set up, apart from the universities, by the Russian Academy of Sciences and that these institutes attract most of the able scientists. He went on to say that it is the belief of most scientists that the traditional association of research with universities in the Western countries is precious. Separate institutes for basic research would drain the universities of their best talent and would deprive the students of contact with the most outstanding scholars. These are the reasons why I have and still do oppose the spending of money to found an institute separate from the universities.

So far I have had in mind a dozen or so of our large universities, most of them state supported. In Canada we have several small private institutions which are called universities. Is there justification for considering these in plans for research development? The answer is, I think, definitely yes. Here is a rich recruiting ground for future research mathematicians. In a survey a few years ago, one small Canadian university stood near the top among all the universities on the continent in regard to the proportion of its graduates who went on to become successful scientists. In these small universities one can successfully steer a large proportion of the talented students away from the less exciting fields and into mathematics, where at present the need is greater. But we cannot capitalize fully on this unless the members of the mathematics departments have a continuing interest in scholarship and research. The compelling reason is that they cannot do the sort of teaching that will encourage students to aim to be ranking mathematicians unless they themselves are advancing knowledge. I think it was Ashbury who said "Universities are traditional places where research is done." This is a very minor reason for requiring university teachers to advance knowledge. The main and compelling reason is that they cannot do the sort of teaching that is required of them unless they themselves are advancing knowledge. In many ways the quiet life of the small college is more conducive to successful research than is the atmosphere of the large urban university. In our small colleges, as well as in our large universities, the mathematics departments should be setting the pace for all other departments. They are under no essential handicap. The other sciences need expensive equipment. The humanities must have elaborate library facilities, not so mathematics. Once the road one wishes to follow is plotted, the library needs are easily supplied by reprints, microfilms, photostatic copies, and inter-library loans and the cost of these to the university is negligible. There is the complaint that in the smaller institutions there is no one to talk to, but this is no alibi. There are numerous conferences and society meetings—far too many, as a matter of fact. It used to be that in the two-day meetings of the American Mathematical Society, one could pick up enough ideas to keep one working for six months. When the American Mathematical Society started taxing the universities and colleges according to the number of papers published over the preceding five years, here in Canada a small college with a registration between five and six hundred tied for first place with our largest university.

As Sylvanus Thompson says in his *Calculus Made Easy*, "what one fool can do another can." What one small college can do another can. What I say is that we in Canada should continue to capitalize on our long summer vacation and all our resources should go into the sectional development of our universities; not a plugged nickel should go into a national institute for mathematics. In all universities, small and large, the mathematics departments should set the pace in scholarships and research.

I. N. SNEDDON

I think the English picture and the Scottish picture are more like the French picture than the Canadian. Mathematical research in Britain is centred in the universities; the Institute in Dublin, after all, is more concerned with mathematical physics than with mathematics, and is very small by American standards. Some of the industrial firms have of course their own laboratories and mathematical departments, but the main location for research in mathematics in Britain is still the universities and it is still in a highly disorganized state. It is very much an individual activity, and springs from the work of individual professors. There has been no attempt, for instance, to co-ordinate things in any way. The financing of research is done entirely through university salaries with no supplements for doing research during the summer months or anything of that kind. It is regarded as a natural activity of a person holding a university post. As far as the young people are concerned, they are financed in the older universities by the universities themselves. Oxford and Cambridge and the Scottish universities, which go back over five hundred years, have large endowments and they have the tradition of giving research scholarships to the ablest people. This has been supplemented on a considerable scale in recent years by the Department of Scientific and Industrial Research which awards maintenance allowances. Expenses are much higher in Oxford and Cambridge, but the grant for average universities is about three hundred and fifty pounds a year. Allowing for cost of living differences, you might equate that to about twelve hundred dollars per year and for Oxford and Cambridge it would be something like sixteen hundred dollars per year. It doesn't seem much and it isn't much but the student can live on it. Also the fees for higher degrees are modest, so the students don't have that problem. One of the things which we feel most, I think, in Britain is the lack of funds for inviting people from abroad to visit England to come and talk to seminars. We are miserably badly off in this respect. If, for instance, an American of distinction happens to be spending a year in Paris, then we can find the money for his trip to London. But he has to be entertained privately! The British Council has a scheme whereby it will put up a certain amount to bring visitors for two weeks from the Western European countries and that is really the extent of the finances we have for this purpose. As far as national meetings are concerned, there are now three activities. The British Mathematical Colloquium meets annually, and now the British Theoretical Mechanics Colloquium, which is really just applied mathematics, also meets annually. They are pretty intensive affairs. Work begins about nine o'clock in the morning and people are still giving talks at ten o'clock at night! They last only about four days but in these four days the packing is much too solid! Perhaps in Scotland we go to the opposite extreme and have a too leisurely affair, lasting ten days, which is really more like your seminar which we have just experienced than the

British Mathematical Colloquium. We try to get people from abroad, and people from England to come and give short courses, and we also have one-hour invited addresses. I think perhaps a mistake was made in holding them in St. Andrews, where there are so many other attractive things to do that it is a bit of a strain for people to do mathematics, but I think that on the whole this is a development which I would rather see than the British Mathematical Colloquium. The more leisurely approach is definitely more pleasant and I think it is more fruitful. And the London Mathematical Society has recently, with the help of some money from UNESCO, organized a two-week school followed by a conference on functional analysis, again very much like the Canadian seminar. In fact, I think that Canada has less to worry about in this respect than any country with which I have had contact. I think that the summer institutes and these biennial seminars are features of Canadian mathematical life which are well known abroad. And, if I may say so, I think that they are decisive factors when a young man in England is trying to make up his mind whether he wants to go and teach in a Canadian university. He knows about the summer institutes and the seminars, and these are most attractive features of the Canadian mathematical scene. It would be terrible if these projects were to end; every effort should be made to continue and expand them.

There are a couple of other remarks that I wanted to make. Some reference has been made to the Soviet Union and their institutes. I made a very brief visit to the Soviet Union and, of course, saw only some things, and some places. But I do know that in these highly specialized institutes of the Academy, it is quite normal for a research worker spending the bulk of his time on research to give one graduate course in the university at either Moscow or Leningrad. They are not entirely separate from the universities. They may be separate organizationally but the contact between them and the senior students is very close. I do not share with Professor Jeffery the fear of an institute. I think that if you had an institute in Kingston it might solve a lot of the difficulties because it would continue the summer institute idea and would be a focus for this kind of activity. I think that an institute which was an ivory tower would be a mistake. You might have to have a plan so that no one could stay longer than two years.

Such an institute would avoid the danger Professor Jeffery has outlined if the smaller universities had the policy of giving generous leaves to members of staff to come to the institute for a short period, say a term or a year, and then return to their own university. If you do not develop this kind of thing, sabbatical years will be taken elsewhere.

I remarked to Hardy about the business of isolation that one of the difficulties of accepting a post in a small university is that you might find yourself doing research in a place where there is nobody you could talk to. He was amazed and said that it would only cost tuppence-halfpenny to send a letter from one part of Britain to another and I replied that he was the last person to talk—he has gone on record that his best work has been done in collaboration with other people. He made the astonishing revelation that he never once had a mathematical conversation with Littlewood! That their being together in Cambridge was an accident; that they wrote to one another, and just dropped a letter in a pigeonhole instead of sending it through the mail. I don't know if there is any value in that observation but I give it to you for what it is worth.

CHAIRMAN

It is true, as you say, that the rest of the world has emulated our activities. The first time I attended a meeting of the Council of the American Mathematical Society, I was asked to give an account of what the Mathematical Congress was doing in Canada. This was the one thing that interested them. And, as you see, in the last few years they have copied many of our ideas. Another case is Australia where they are making every effort to copy and benefit from what is going on in Canada. I wonder if Professor Erdélyi would say something about American reaction to what we have done.

A. ERDÉLYI

Mr. Chairman, I am really ill prepared to give any sort of general summary of the situation. Moreover, I do feel that at least half of the people sitting in the audience know as much about the American mathematical situation as I do. With your permission, I would like to restrict myself to a few marginal remarks. I too do not see eye to eye with Professor Jeffery on the question of a central institute. As Professor Sneddon pointed out, in Russia and also in other European countries, a person attached to a research institute is not lost to the universities. Far from it. And even Princeton, which looks the most ivory of ivory towers in mathematics, is not altogether isolated from university life. Consider how small the number of permanent members of the Institute is and how large the number of those who come, as Professor Sneddon suggests, for a semester, for a year, or for two years, either from American universities or from Europe, who would otherwise never speak or write to each other. They derive a great deal of stimulus from their sojourn at the Institute and go home to brighten large areas of American university life. So much for the Institute. Now I would like to make two points with some of the exaggeration that is occasionally heard from this place. One is this: I fully agree with Professor Jeffery in saying that the great importance of research in the universities, by university staff, is due, at least to some extent, to the fact that the kind of teaching that should be done in the universities cannot successfully be performed except by people who take part in developing mathematics. From this point of view, I would like to make a clear distinction between research and publication. To some extent, I feel that these two activities are opposed to each other, curiously enough. As long as promotion in universities depends on the number of papers, or the total number of pages one publishes, the pressure is heavy, especially on younger people, to publish a great deal of trivial stuff rather than to engage in significant research. I know that, from the point of view of the departmental chairman, publications are the easiest means by which to convince an administration that this fellow deserves promotion. But among ourselves, we are perfectly aware of the practical difficulty involved in this. I do feel that if we encourage research in universities mainly in order to attract the right type of person and to enable him to transmit mathematics in the way he feels he should transmit it, then if such a person sits down, thinks for two years, and discovers Poincaré's last theorem, he is much more able to transmit his joy in mathematics and his appreciation of creative mathematics than if he proves a formula or a theorem that nobody has bothered writing down before and which, therefore, modern periodicals will accept. I would not like to go so far as to prohibit or discourage publication but I would like to put a premium on good presentation, or good style, rather than on novelty in the technical sense and on the amount of publication.

The second recommendation I should like to make to all universities who can afford it is derived from our own experiences at the California Institute of Technology. In many American universities (I don't know whether this is true in Canada) it is said with truth that older staff members have a great deal of university responsibility—they are on committees, they take part in running the mathematics department—but a young instructor has none. Therefore, a young instructor can be asked to teach far more than his elders. I know universities, and very good universities, excellent mathematics departments let me say, where an instructor teaches from nine to eleven hours when a full professor can get away with three hours of teaching. We have inflexibly followed the policy of allotting approximately the same number of hours to all members of our staff. As a matter of fact, we are easier on instructors than on full professors. If they want to have maximum time for their research, we often give them two undergraduate classes when a full professor usually teaches one undergraduate class and one graduate class, the latter very often a course which he is writing during the year when he gives it. We do this because we feel that it is the young man, the one who is working himself into a field, who needs most time for his research. We feel that giving a young man a very heavy teaching load, and asking him to produce papers or face the penalty of slow promotion, means that we practically push him into working out further and further special cases, and little side issues to his doctoral thesis instead of giving him time to do a great deal of miscellaneous reading, to find the field of mathematics for which he is best suited and in which he is most interested and to engage in a research project of considerable magnitude.

C. Fox

I wish to address a question to Professor Erdélyi. Again and again we see papers in American periodicals, with "this was supported by a grant from the Navy, number xyz" and so on. My question is: Do you think this is desirable or not?

A. ERDÉLYI

Yes and no. There are really two kinds of support which the armed forces give to mathematical research. They support the research by enabling universities to give what is virtually a post-doctoral fellowship to a young man, to invite a distinguished mathematician from abroad or from another university. This is the part to which my answer is an emphatic yes. This is a very fruitful and a very useful way of stimulating mathematical research. It is felt that in the course of time this function will largely be taken over by the National Science Foundation. There is, however, another way in which these grants support research at which I frown on to some extent. In most American universities one gets paid for nine months only, the theory formerly being that for the remaining three months if you had private means you travelled or did whatever you liked and if you had not you got a job as a dish-washer or teacher at a summer school. I do not like this, especially if we feel that universities are natural homes of research as well as teaching institutions. It follows that university staff should be paid for twelve months even if they do not teach summer school. I understand that in most, if not in all Canadian universities this is indeed the case, as it is, I am happy to say, at the California Institute of Technology. We work on a system very closely resembling that of European universities. We have a solid three-months' vacation. There is no summer school at all at the Institute. If I want to do research work in Pasadena, I do it on my salary. If I want to go to Europe I get my salary, and if I happen to be working on a project sponsored by the Office of Naval Research, I still get the same salary. In some cases the Institute does get a refund for part of it and in some cases it does not, but I do not notice the difference. To me this is both a dignified and a desirable arrangement, the sort of arrangement towards which American universities should strive. But I fear that the availability of support for research from sources outside the university—the possibility of good men doing research instead of dishwashing or teaching summer school and still getting a salary—has contributed to lessening the pressure on university administrations to adopt this enlightened attitude. In so far as it has done this, I find the support less than desirable.

L. HENKIN

I do not have a recommendation to bring before you but I have been struck by one omission in the discussion of Canadian policy for stimulating mathematical research. That is the fact that the talk has been in terms of whether or not an institute should be established or whether there should be this kind of exchange between Canadian universities or that. I well appreciate even in my short stay here that the desire of Canadians, in all walks of life, is to maintain their individuality as a nation. Nevertheless, it seems to me that scientists and mathematicians should not be bound closely by national interests but that some attention should be given to the fact that mathematics is, after all, a universal activity rather than the property of this or that country. It seems to me that it would be unfortunate for both our countries if whatever policy was worked out failed to take advantage of our proximity and the ease of moving back and forth between our two countries. I would hope that your efforts would not be directed entirely to keeping Canadian mathematicians in Canada during the summer and that we would be able to see all of you in our American universities and that we would be afforded the opportunity of coming here as we have been so generously invited this time.

CHAIRMAN

I think that Professor Henkin, living in California, perhaps does not realize the movement between the two countries which already exists. In fact, before the war, the pull to the south was so strong in Canada that we did not know other Canadian mathematicians at all. However, the aim of the Congress has not been to compete with the American Mathematical Society but to supplement it. We have always had the feeling that this is part of our life, that the short paper sessions of the American Mathematical Society bulk large on our horizon. I think this is the way to look at the problem.

M. L'ABBÉ

I would like to make one point which has not been raised. Mathematical research is usually done in universities but not always by permanent staff members. In Canada, almost the only source for postdoctoral research fellowships is the National Research Council, which awards not more than ten in mathematics whereas many more are available in experimental science in the universities and in the National Research Council itself. I think we should try to convince the Council that this number in mathematics should be increased. I believe that to have the means of inviting foreigners or professors from other parts of Canada to come to our universities to spend a year or a part of a year is a very good way to promote research. If we had more money we could invite people for short conferences or colloquia; we have, of course, a biennial seminar that lasts three weeks—a long stretch, especially for those who are spending three months at the Summer Research Institute. I think that Canada is large enough that we should be able to provide for small colloquia in various parts of the country during the academic year from money made available not by the Congress, but by provincial agencies.

THE TEACHING OF APPLIED MATHEMATICS IN CANADA*

A. F. PILLOW, *University of Toronto*

I should like to discuss four topics. First, since there is a range of opinions, I had better indicate roughly what I understand by applied mathematics. Secondly, I want to discuss what type of training an applied mathematician requires. Thirdly, I should like to make some detailed suggestions on a syllabus for an undergraduate course in applied mathematics. Finally, I should like to talk about what can be done at the graduate level.

WHAT IS APPLIED MATHEMATICS?

It seems worth while to me to attempt a classification of the theoreticians, who use mathematics to a significant extent, into four different classes which we might call: (*a*) *the pure mathematicians*—those concerned with the analysis of abstract structures and with the doing of algebra, geometry and analysis; (*b*) *the theoretical scientists*—those concerned with the construction and understanding of idealized models which fit the scientific facts either in the exact sciences, the biological sciences, or the social sciences; (*c*) *the theoretical engineers*—those concerned with the design, understanding, and maintenance of the constructs of applied science; and (*d*) *the applied mathematicians*—those concerned with the deductive analysis of the accepted idealized models of science and engineering. In this classification, the statistician would fall partly in (*a*) and partly in (*d*).

The above are not watertight divisions and I think there should be free interplay between all four. I do believe, however, that for didactic purposes at least three different types of mathematics courses are needed: (i) pure courses, (ii) applied courses, and (iii) service courses, and in fact, four, if we separate statistics. An

*Presented at panel discussion on Applied Mathematics at the Canadian Mathematical Congress, September 5, 1961.

additional course may be needed if we consider the needs of school-teachers.

Among the (*b*) and (*c*) groups are many who need a strong knowledge of conventional processes in mathematics—at the very least up to a reading level. For these people, then, there must be a service course available, taught by mathematicians with some practical experience of the mathematics needed and with an ability to exemplify it in the relevant field.

WHAT TYPE OF TRAINING DOES AN APPLIED MATHEMATICIAN REQUIRE?

For the middle man, the applied mathematician, something more is needed, for he is responsible for pushing the analysis of physical models as far as mathematics will allow. He must then be solidly trained in the creative use of mathematical methods. This is his biggest threshold and I think he should, therefore, be housed with mathematicians. However, since he will not primarily be analysing new mathematical structures, he does not need all the intensive abstract training of the pure mathematicians.

The second vital feature of an applied mathematician's training is, I feel, that he must *from the very beginning* be given adequate training in the deductive treatment of the classical models of science. The most suitable people to teach such courses are those mathematicians who through their research have gained some sympathy with and experience of the model-making side of applied mathematics.

It should be emphasized that there is no great clash or overlap of interests here with the scientist. For the interest is not so much in unravelling the causes at work in the new phenomenon but rather in deducing in detail, over a range of examples, the finer consequences of the phenomenon from the basic model originally provided by the scientist. Once this is achieved, the way is open mathematically for adding further features to the model at the request of the scientist who ranges more widely over the whole field of his subject. Usually, the applied mathematician can only afford a detailed knowledge of the scientific facts in his own applied field. Perhaps one can say that the applied mathematician builds up to the actual problem synthetically from mathematical sub-models, while the scientist analyses the actual physical situations into their basic phenomena.

If the syllabus is properly graded, it is certainly possible to teach applied along with pure and the relevant sciences in the first year at the university. Many universities in Britain and Australia do just this. The interplay of mathematical method with scientific intuition is healthy and by no means a one-way process. Point sources and multipoles were understood long before distributions and their derivatives were thought of. To postpone the treatment of applied mathematics to the last years of a mathematics course is as absurd as insisting that geometry cannot be studied until the algebra of transformations and their invariants is mastered.

Before the synthetic model-making side of applied can be taught, it is necessary that a sufficient arsenal of elementary standard models be built up and analysed mathematically. In the physical sciences, this takes at least three years and means that a start on applied is imperative in first year if the synthetic side is even to be indicated before graduation.

As well as the pure and applied courses, it is important that the student attend the relevant science courses and gain a grasp of the scientific facts and experience of real world situations. This can certainly be achieved without overloading the time-table if the scientists are in sympathy and the student is allowed to forego some of the practical work, particularly in the later years.

AN UNDERGRADUATE COURSE IN APPLIED MATHEMATICS

In Canada, it appears that almost no full-blown undergraduate applied mathematics courses are available. To me, it seems that the responsibility for sponsoring such courses lies squarely with the mathematics departments, since the bulk of the teaching calls for a solid mathematical training. To most scientists, the objectives of an applied course appear as a side issue. Indeed, much of classical physics can be formulated and understood in principle without too much advanced mathematics. The result has been that applied courses have not gained a hold scientifically even in physics departments, and that work on the mathematical aspects of classical physics has languished even though there is much for the mathematicians still to do. The physicists, understandably, have concentrated on training their students for the exploration of new fields of physics rather than for the detailed deductive analysis of the old, well-proven models of the classical branches of their subject.

Up to the present, the major fields of application of mathematics have been in the physical sciences. If we are to build up applied courses with a common core in several universities, then these sciences seem the natural starting point.

To come down to details, let me describe, as an example of what might be done, the undergraduate course in applied which we are hoping to build in Toronto. We are fortunate in inheriting from the heyday of Synge and Infeld a time-table in which applied mathematics is given recognition. Two hours per week are available in the second year, four hours per week in the third year, and seven two-hour periods per week in the fourth year. We are hopeful that a further two hours per week course will be introduced in the first year. If this is achieved, we feel it will be possible to give an introductory three-year course which avoids specialization and which strikes a balance between classical dynamics and the simpler scalar and vector field theories of physics. We hope this will kill the false bogey that applied mathematics is just difficult dynamics—a reputation which the old Cambridge Tripos helped to establish.

In the first year, dynamics of particles, systems of particles, and rigid bodies would be taken vectorially in one and two dimensions mostly. In the second year, three-dimensional dynamics, Lagrange's equations, and vibrating systems would be discussed briefly and in the remaining time, simple scalar fields such as the temperature field in heat conduction would be treated. In the two third-year courses, the elementary models in fluid dynamics, electromagnetism, and elasticity would be treated along with the relevant potential theory, vector field theory, and the use of complex variable methods.

The fourth year would be devoted to those wishing to specialize in some branch or branches of applied mathematics. The topics offered at present are (*a*) generalized dynamics and quantum theory; (*b*) tensor calculus and relativity; (*c*) fluid dynamics; (*d*) elasticity; (*e*) partial differential equations; (*f*) electromagnetism; and (*g*) numerical analysis. Of these, four courses would be compulsory for applied mathematicians who would choose their remaining three courses from pure, statistics, physics, or the remaining applied courses.

The allotment of time during the first three years would be roughly as follows in units of "hours per week."

First Year: Common to mathematics, physics, and chemistry

	pure	5	chemistry	6
	applied	2	cultural	3
	physics	7		

Second Year: (pure, applied, and statistics)

	pure	6	physics	6
	applied	2	astronomy	1
	statistics	1	cultural	1

A special service course of only 3 hours per week in pure is offered to scientists.

Third Year: (applied)

	pure	6	optional (pure or statistics)	4
	applied	4		
	physics or economics	2		

In the third year, one-half of the applied is optional for the pure and statistics students. The scientists are given a separate service course of two hours per week in pure and they do one of the applied courses. There are also special pure service courses spread over several years available to students of engineering physics.

With the schedule I have just given, the applied mathematicians neglect neither the pure side nor the physics side. At the end of the fourth year, an applied student would have had lectures on at least the following pure topics: real and complex variable analysis and special functions; linear and tensor algebra; ordinary and partial differential equations; analytical and differential geometry; introductory modern algebra; elementary topology; probability theory and statistics. Much more pure and statistics would be available to those who specialized in only a limited number of branches of applied.

GRADUATE WORK

The lack of adequate undergraduate courses in applied has resulted in a critical shortage of both staff and graduate students in the applied field in Canada. To alleviate this shortage quickly, the only practical solution appears to be for each university to specialize in one aspect of applied at the graduate level. To begin with, it would be sufficient if, say, five universities in Canada were to do this. Obtaining staff (from overseas if necessary) should not then be impossible. This is the solution we are adopting in Toronto where we have persuaded five young fluid dynamicists to join us. Of the ten applied members of staff there will be seven at or near the postdoctoral level working in this field next year. I hope that this group will provide a sufficiently stable, concentrated nucleus to attract research students.

One major difficulty is that graduate students who come to us to do research in applied have first to be trained in elementary undergraduate applied, and much valuable time is lost. This waste could be avoided if the universities offering graduate work in applied were to offer a broad, balanced undergraduate applied programme along with the pure and statistics courses. It would then be posssible for students to transfer from one university to another for their graduate work. Thus the development of specialized graduate schools and a healthy cross-fertilization would be possible.

Research training is a vital part of the teaching of applied in universities. It is only at this level that a proper appreciation of the rôle of synthetic model-making in applied can be developed and an understanding gained of the richness of interplay between approximation in the model and the success of the mathematical analysis.

CONCLUSIONS

There is a dire need for the establishment of a common four-year undergraduate applied mathematics course in the mathematics departments of at least five Canadian universities.

About four or five specialized graduate schools are needed to begin the training of academic staff and research workers, which we cannot expect to draw continually from outside Canada.

If these conclusions be acceptable, then I feel the Congress, along with its members, should do what it can to bring this educational gap to the attention of university authorities.

ABSTRACTS OF CONTRIBUTED PAPERS

ABSTRACTS OF CONTRIBUTED PAPERS

ON CONSERVATIVE DYNAMICAL SYSTEMS AND THE EQUATION OF WAVE MECHANICS

NICHOLAS CHAKO,
Queens College and Institute of Space Studies, New York

The purpose of this paper is to examine more closely the conditions which a classical dynamical system must satisfy in order to derive a corresponding wave equation associated with such systems. Here a method is developed by which the equations of wave mechanics (Schrödinger, Klein–Gordon, Dirac) are derived from the Hamilton-Jacobi equation based on certain properties of the classical system and the Einstein energy-frequency relation. These properties are connected with the fundamental theorem of Poincaré pertaining to the existence of uniform integrals and the stability conditions of dynamical systems. The results show that the characteristic exponents of Poincaré play a fundamental rôle in our procedure for deriving an associative-wave equation corresponding to a classical dynamical system. Apart from the fundamental hypothesis of Einstein, the arguments presented here are based solely on classical considerations of a dynamical system. Furthermore, these arguments can easily be carried over to optics.

MULTIPLICATIVE IDEAL THEORY AND RINGS OF QUOTIENTS

G. D. FINDLAY *and* J. LAMBEK, *McGill University, Montreal*

An ideal A of an associative ring R with identity is called *invertible* in an over-ring S of R if and only if there exists a submodule B of S such that $AB = R = BA$. We show that A is invertible in some over-ring S if and only if it is invertible in the largest ring Q of right and left quotients of R. We call the ideal A of R *integral* if all ideals of R containing A are invertible. Multiplication of integral ideals turns out to be commutative, and the unique factorization theorem holds for integral ideals. To look at a classical example, let R be an integral domain, S its integral closure, and assume that S is finitely generated over R and that it is a Dedekind domain. Then the integral ideals of R are precisely the ideals of R which are prime to the conductor $R : S = \{q \in Q : qS \subset R\}$. In the general situation we also introduce a so-called conductor with the property that all ideals prime to it are invertible.

We have also extended the Van der Waerden-Artin-Pruefer theory to arbitrary associative rings with identity. An ideal A is called a *p-ideal* if, for all ideals B containing A and all elements q of Q, $qB \subset B$ or $Bq \subset B$ implies $q \in R$, and furthermore $A = C^{-1} = \{q \in Q : qC \subset R\}$ for some submodule C of Q. If A and B are p-ideals so is their *p-product* $(AB)^p$, which is defined as the smallest p-ideal containing AB. The Artin refinement theorem holds for p-ideals under p-multiplication. A p-ideal is called an *m-ideal* if the p-ideals containing it satisfy the maximal condition. The unique factorization theorem holds for m-ideals under p-multiplication, which is commutative for m-ideals.

P-ALGEBRAS AND DERIVATIONS

K. HOECHSMANN, *University of Notre Dame, South Bend, Indiana*

Let K be a field of characteristic p with an extension Z such that $Z^p \subseteq K$ and $(Z\colon K) = p^m$. Take a derivation d of Z into Z such that $dz = 0$ if and only if $z \in K$. Then the K-algebra (Z, d, γ) generated by Z and a symbol u with relations $uz = zu + dz$ and $f(u) = \gamma \in K$, where f is the minimal polynomial of d, is a centrally simple algebra containing Z as maximal subfield. Conversely, every such centrally simple algebra is of the form (Z, d, γ). There is a homomorphism V from $(Z, +)$ into $(K, +)$, such that (Z, d, γ) is isomorphic to (Z, d, γ') if and only if $\gamma - \gamma' \in V(Z)$. (Z, d, γ) splits if and only if $\gamma \in V(Z)$. Finally, $(Z, d, \gamma) \otimes (Z, d, \gamma')$ is similar (in the sense of Brauer) to $(Z, d, \gamma + \gamma')$. Evidently the theory of these algebras is formally similar to that of cyclic algebras.

The following generalization of this construction can be used to build up p-algebras recursively. Given a centrally simple Z-algebra A, we can extend d to a derivation d' of A, and obtain an element $\alpha \in A$ such that $\alpha a = a\alpha + f(d')a$ for all $a \in A$ and $d'(\alpha) = 0$. Then the K-algebra (A, d, α') generated by A and u such that $ua = au + d'a$ and $f(u) = \alpha$, is centrally simple and contains A as centralizer of Z. Conversely, every such K-algebra is given by $(A, d', \alpha + \gamma)$ with $\gamma \in K$.

FLOW OF HEAT FROM A WEDGE INTO A SURROUNDING MEDIUM

P. MANDL, *National Research Council, Ottawa*

This paper, which is a condensed version of a report of the National Research Council of Canada **(1)**, presents the calculation of the transient temperature field due to the cooling of a thin, liquid wedge by a surrounding infinite solid.

The main object of the investigation is to determine the motion of the isothermal surfaces and of the solid-liquid interface formed during the cooling process. Since the interface is a moving surface whose position is unknown and at which boundary conditions must be satisfied, the problem is one of great difficulty; other complications are due to the media inside and outside the wedge being different. Therefore the problem is that of conduction of heat in composite materials which has been solved only in the case where the temperature depends on only one space variable, changes in phase being ignored **(2)**.

A solution by a method of successive approximation is here proposed and analysed. The solution to the first approximation, in which changes in phase are ignored, requires, in essence, extending one-dimensional theory of conduction of heat in composite media to two dimensions. It is shown that the equation of heat-conduction and the boundary conditions admit a self-similar solution, which is a function of only two independent variables, namely, the angular co-ordinate and a combination of radial co-ordinate, time, and diffusivity.

The solution to the first approximation, in turn, must be determined by a step-by-step procedure. It seems legitimate to assume that, in the case of a thin wedge, the influence of the transverse temperature gradients is more important than that of the radial gradients. This suggests as a first step to ignore the radial gradients in the equations. The solution to this simplified problem can be represented by two types of infinite series which complement each other: one, involving error complementary functions, the other ortho-normal functions. At large distances from the vertex, the series represent one-dimensional heat flow, the lines of flow being circles. This approximate solution may now be improved by iterative calculations taking account of the influence of the radial gradients. This leads to an infinite number of inhomogeneous boundary value problems which are solved successively by the technique of the Green's function.

The theory is then applied to the practical case of the diffusion of heat due to a wedge-shaped molten intrusion into solid granite. The simpler theory, which ignores the influence of the radial gradients, predicts the motion of the isothermal surfaces to a high degree of accuracy everywhere, except near the vertex, where the influence of the radial gradients becomes important. It is interesting to observe that all isothermal surfaces possess, as a common asymptote, the face of the wedge which remains at a constant temperature; also, the isothermal surfaces are seen to recede from the vertex at a decreasing rate.

In an appendix to this discussion the theory is extended to take into account the change in phase of the medium. The preceding calculations provide, then, the approximate position of the isothermal surfaces, and, in particular, that of the interface, corresponding to the temperature of solidification of the liquid. These approximate results may be considered as a starting point for calculating a better approximation which takes account of the latent heat of fusion and the different thermal constants describing the two-phase system on either side of the interface. This new cycle of calculations, in turn, yields corrections to the motion of the isothermal surfaces and thus the interface.

It is interesting to note that the concept of a self-similar solution applies even when latent heat of fusion is included in the formulation. However, owing to increased complexity, this extended theory has not been applied to a numerical example.

REFERENCES

1. P. Mandl, Flow of heat from a wedge into a surrounding medium, N.R.C. Aero. report, LR-312.
2. H. S. Carslaw and J. C. Jaeger, Conduction of heat in solids, 2nd ed., Oxford (1959).

HOMOMORPHISM BETWEEN GROUPS OF CONTACT TRANSFORMATIONS AND CORRESPONDING GROUPS OF POINT TRANSFORMATIONS

JOHN MARIANI, *Rutgers University*

There is between some classes of groups of contact transformations and corresponding classes of continuous groups of point transformations a homomorphism which has not been pointed out in the classical Lie theory of groups. In the (x, y)-plane, there are three classes of primitive groups of point transformations. Class I defines the special linear group, class II the general linear group, class III the general projective group. On the other hand, in the (x, y, z)-plane there are four classes of groups of contact transformations. Class I defined by the differential operators* $p, q + xr, r, xq + \frac{1}{2}x^2r, xp - yq, yp + \frac{1}{2}y^2r$ is the homomorphic image of point transformations of class I in the (x, y)-plane. Class II with the operators $p, q + xr, r, xq + \frac{1}{2}x^2r, xp - yq, yp + \frac{1}{2}y^2r, xp + yq + 2zr$ is homomorphic to class II of point transformations. There is no class III of contact transformations and class III of point transformations cannot be homomorphic to any group of contact transformations. Class IV of contact transformations defined by the operators* $P - YR, Q + XR, R, XQ, XP - YQ, YP, XP + YQ + 2ZR, ZP - Y(XP + YQ + ZR), ZQ + X(XP + YQ + ZR)$ is associated with the conformal group of the (x, y, z)-space defined by the operators* $p, q, r, xq - yp, yr - zq, zp - xr, U, 2xU - Sp, 2yU - Sq, 2xU - Sr$, by means of Lie's contact transformation changing

*$p = \partial/\partial x$, $9 = \partial/\partial y$, $r = \partial/\partial z$, $P = \partial/X$, $Q = \partial/Y$, $R = \partial/Z$, $U = xp + y9 + zr$, $S = x^2 + y^2 + z^2$.

spheres into straight lines and asymptotic lines into curvature lines. Moreover, reflections in the (x, y)-plane are homomorphic images of finite contact transformations.

A Note on Weighted Compositions

T. V. NARAYANA, *University of Alberta, Edmonton*

A simple direct proof of the following theorem of Moser and Whitney **(1)** is given: The number of compositions of n into parts $\equiv 1, 2, 4$, or 5 (mod 6) and involving an even number of parts $\equiv 4$ or 5 (mod 6) exceeds by n the number of compositions of n into parts $\equiv 1, 2, 4$, or 5 (mod 6) and involving an odd number of parts $\equiv 4$ or 5 (mod 6). The difference equations, $g(n) = g(n-1) + g(n-2) + \ldots \pm g(n-r)$, which provide a generalization of the Fibonacci series as well as of the above theorem, are being studied.

Reference

1. L. Moser and E. L. Whitney, Can. Math. Bull. *4*, 1 (1961).

The Dynamics of the Expanding Vapour Bubble in a Boiling Liquid

P. SAVIC, *Division of Mechanical Engineering, National Research Council of Canada, Ottawa, and* J. W. GOSNELL, *Supervisor, MIT Reactor, Department of Nuclear Engineering, Massachusetts Institute of Technology, Cambridge, Mass.*

The theory of a constant pressure vapour bubble growing in a liquid with both superheat and subcooling is presented. The bubble is assumed spherical and in the first part of the paper the initial and subsequent temperature profile is also of spherical symmetry. The conduction–convection equation for this case is solved under the postulate of dynamic similarity and it is shown that this can be maintained only if the initial temperature distribution is uniform. The independent similarity variable is $s = y/a$, where y is the radial distance of a point from the surface of the bubble and a is the bubble radius. To extend the treatment to the case of a non-uniform temperature field, the system is perturbed with a as the perturbation parameter. Thus, the initial temperature T is developed in a power series in $y = s.a$ and $a.(da/dt)$ in a series in a. Equating equal powers of a yields a system of ordinary second order differential equations for the coefficients of the T series. Since the solutions of this system are unknown transcendental functions, a boundary layer approach was adopted, wherein the known asymptotic solution for large s is joined to the (also known) solution for small s by suitable continuity conditions in the intermediate range. It is shown that this procedure is admissible only if the ratio of liquid enthalpy to vapourization enthalpy (Jakob number) is large. Hence, the Jakob number assumes a significance similar to the Reynolds number in conventional boundary layer theory. Once the coefficients of the $a.(da/dt)$ series are calculated, the bubble growth function can be determined and from this follows the temperature distribution around the bubble at all times after nucleation. It is shown that the bubble growth curve can be made to agree with a numerical treatment by Griffith **(1)** of a hemispherical bubble growing on a hot solid surface, provided the inverse thickness of the superheated liquid layer is averaged over the surface of the bubble.

In the second part the treatment is extended to the case considered by Griffith. This implies that the initial temperature distribution in the liquid now depends both on the radius vector and on the polar angle in a determinate manner. The isothermals in the liquid at $t = 0$ are planes parallel to the heater surface. The similarity approach of the first part is again employed, but now the system of equations resulting is partial, allowing solution by separation of variables. Functions of the radius vector are of a type similar to those of the first part (confluent hypergeometric functions), while the functions of the polar angle are resolved in terms of Legendre polynomials. The reason for the averaging process referred to above now appears justified. The bubble growth law presents nothing essentially new. Similarly, the temperature distribution in the liquid surrounding the expanding bubble follows closely the one for the simpler centrally symmetrical problem, provided the Jakob number is large. For values of the Jakob number of the order unity, however, new features appear. While the liquid near the base of the hemispherical bubble continues to vapourize throughout the life of the bubble, condensation begins to take place near its peak where it penetrates farthest into the subcooled bulk liquid. A situation may be reached whereby as much liquid is recondensed at the peak as is vapourized at the foot.

As the bubble expands from the nucleation centre, it quickly surrounds itself with a thermal boundary layer of almost perfect spherical symmetry, provided the Jakob number is large. For moderate values of this parameter, the isothermal corresponding to saturation temperature approaches the bubble surface and finally intersects it, forming a branch circle. This circle separates the regions of vapourization and condensation. The maximum bubble radius is the root of a polynomial the coefficients of which are related to the coefficients of the power series expansion of the initial temperature profile.

REFERENCE

1. P. Griffith, Trans. A.S.M.E., *80* (1958), 721.

SOME METHODS FOR COMPUTING MODEL STELLAR ATMOSPHERES

ANNE B. UNDERHILL,
Dominion Astrophysical Observatory, Victoria, British Columbia

One major aim of astrophysics is to understand the physical constitution of stars. The general procedure that is followed is to compare the results of theoretical studies with observed quantities. The theoretical representation of the interaction between atoms and radiation under stellar conditions results in certain differential-integro equations with coefficients which are dependent in a complicated way on the independent variable. Such equations like those representing most real physical situations, cannot be solved with sufficient rigour by analytical methods. Recourse must be had to numerical solutions.

This paper summarizes briefly the arguments leading to the equations which must be solved in the case of stellar atmospheres. Summaries are presented (*a*) of the method of solution which has been developed for a modest, moderate-speed, magnetic drum computer, and (*b*) of more powerful methods that depend upon the use of a very fast computer which has a large, rapid-access memory. The problem of solving the equations of radiative transfer is broken down to the problem of evaluating a series expansion for the so-called source function. A few remarks are made regarding the choice of the best expansion, considering the characteristics of the problem.

ATMOSPHERIC WAVES PRODUCED BY A LARGE EXPLOSION IN THE ATMOSPHERE*

V. H. WESTON,
Department of Electrical Engineering, University of Michigan

The pressure pulse produced by a large explosion in the atmosphere is investigated. The explosion is represented in terms of the excess pressure and normal velocity on a closed surface, outside of which the hydrodynamical equations are linearized. The pulse is represented in terms of a Fourier transform of the associated harmonic frequency problem, for which a ring source Green's function is obtained in terms of an expansion of the discrete modes. The distribution of the discrete spectrum is discussed for two general models of the atmosphere. It is shown that the excess pressure may be represented in terms of an integral (containing the Green's function) over the surface surrounding the source. A point source representation for large explosions is then used to simplify the expression for the portion of the pressure pulse directly received.

*The research reported in this paper was supported by the Air Force Cambridge Research Laboratory, Geophysics Research Directorate, Office of Aerospace Research, United States Air Force, under Contract AF 19(604)-5470.

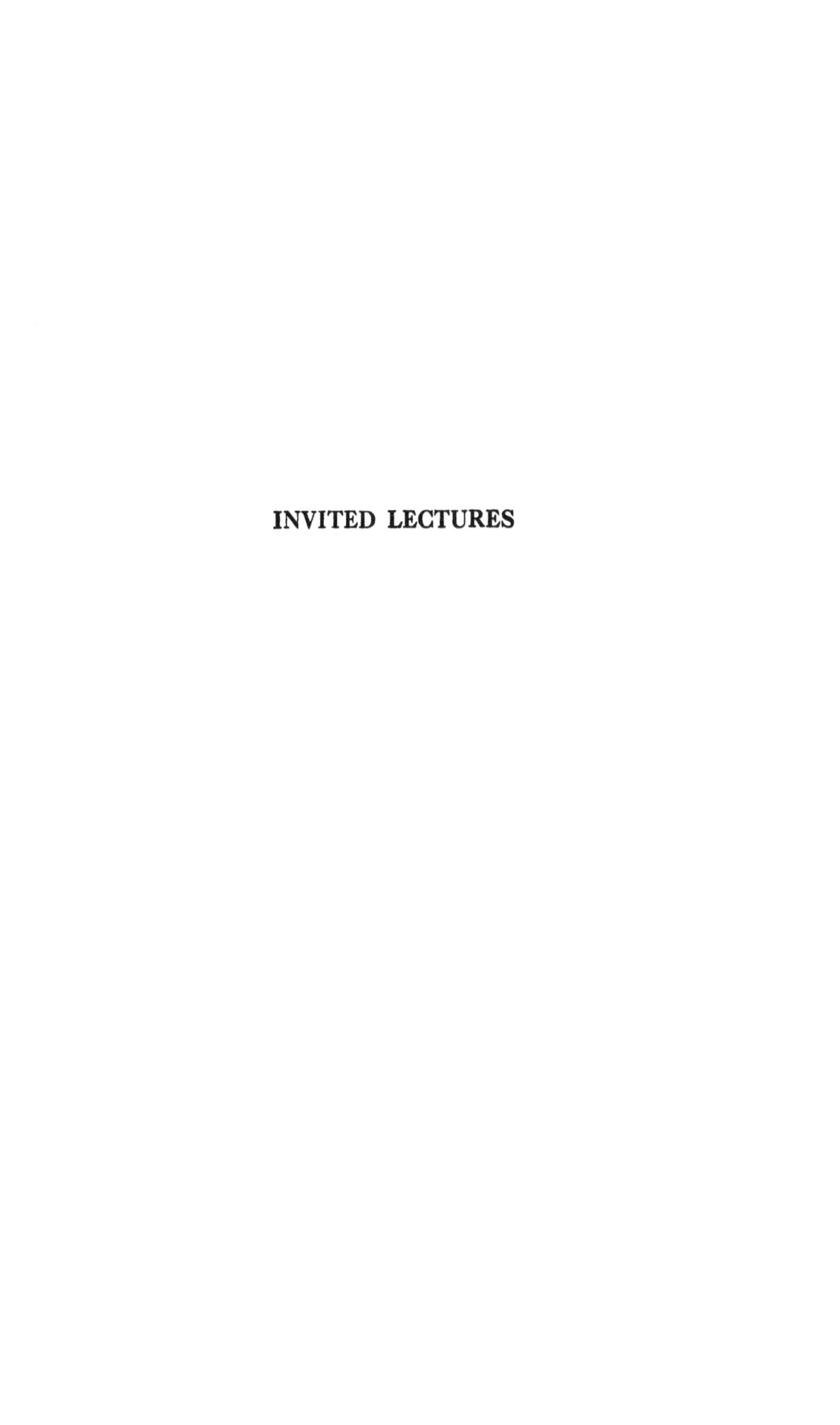

INVITED LECTURES

STRUCTURES FEUILLETEES

CHARLES EHRESMANN, *Institut Henri Poincaré*

Introduction

Cet article a pour but la définition précise et l'étude des structures feuilletées dans le cadre de la théorie des structures locales telle qu'elle est exposée dans **(3; 5; 6)**. Les résultats connus dans le cas des variétés feuilletées sont précisés et généralisés au cas d'un feuilletage topologique localement simple. Les notions d'holonomie, de déroulements et de tubes analysées ici permettent d'étudier les questions de stabilité. Seuls des problèmes généraux sont abordés, les applications étant réservées pour une publication ultérieure.

La plupart des idées contenues dans ce travail ont été exposées dans mes cours (en particulier, Paris 1955–56, 1958, 1961) et dans des conférences (par exemple Princeton 1953, Buenos-Aires 1959–60, Montréal 1961). Rappelons que la notion de variété feuilletée a été introduite dans une Note en collaboration avec Reeb **(1)**, puis étudiée d'une façon approfondie par Reeb **(13; 14)** dans différentes publications. Les structures feuilletées d'espèce $\mathfrak{B} \# \mathfrak{F}$ élargie et de seconde espèce ont été définies dans **(3)**. Les Γ-structures étudiées par Haefliger **(16)**, qui sont étroitement liées aux feuilletages de seconde espèce, ne seront pas considérées ici. Les feuilletages localement simples ont été introduits dans une Note en collaboration avec Shih Weishu **(2)**.

I. Définitions de diverses espèces de structures feuilletées

1. *Feuilletages topologiques*

Soit E un ensemble muni de deux topologies T et T'. On dira que (T, T') définit sur E un *feuilletage topologique* ou une structure d'*espace feuilleté topologique* si la condition suivante est vérifiée : Pour tout $x \in E$, il existe un voisinage ouvert U' de x relativement à T' sur lequel T et T' induisent la même topologie.

Si (T, T') est un feuilletage topologique sur E, alors T' est une topologie plus fine que T. Nous supposerons désormais que T' est localement connexe. Une composante connexe F de E relativement à T', munie de la topologie induite par T', sera appelée une *feuille* de (T, T'). Si T et T' induisent la même topologie sur F, on dira que F est une *feuille propre*. Les feuilles de (T, T') forment une partition de E; la relation d'équivalence correspondante sera appelée *relation d'équivalence sous-jacente à* (T, T').

Remarque. Un feuilletage topologique (T, T') tel que $T \neq T'$ peut n'avoir qu'une seule feuille. Par exemple, considérons sur le tore à deux dimensions une géodésique E partout dense. Soient T' la topologie de variété à une dimension de E et T la topologie induite sur E par celle du tore. Alors (T, T') est un feuilletage topologique dont E est la seule feuille.

Soit $E^{\vee}$ l'espace quotient de l'espace E, muni de la topologie T, par la relation d'équivalence sous-jacente à (T, T'); l'espace topologique $E^{\vee}$ sera appelé *espace transverse* de (T, T').

Si U est un ouvert de E relativement à T, le couple des topologies induites par T et T' sur U définit un feuilletage topologique sur U, que nous appellerons *feuilletage induit par* (T, T') *sur* U et que nous noterons $(T, T')_U$.

Soient (T, T') et (T_1, T_1') deux feuilletages topologiques sur E et E_1 respectivement. Une application f de E sur E_1 est un *isomorphisme* de (T, T') sur (T_1, T_1') si f est un homéomorphisme de T sur T_1 et de T' sur T_1'. Un *isomorphisme local* de E sur E_1 est un isomorphisme de $(T, T')_U$ sur $(T_1, T_1')_{U_1}$, où U et U_1 sont des ouverts de E et E_1 relativement à T et T_1 respectivement. Soit $E^{\wedge}$ un espace étalé par p sur E; alors il existe un feuilletage $(T^{\wedge}, T^{\wedge\prime})$ sur $E^{\wedge}$ tel que la restriction $p_{U^{\wedge}}$ de p à $U^{\wedge}$ soit un isomorphisme de $(T^{\wedge}, T^{\wedge\prime})_{U^{\wedge}}$ sur $(T, T')_{p(U^{\wedge})}$ pour tout ouvert $U^{\wedge}$ tel que $p_{U^{\wedge}}$ soit un homéomorphisme sur $p(U^{\wedge})$.

Une feuille F de (T, T') sera dite *simple* si tout point de F a un système fondamental de voisinages U ouverts relativement à T et tels que l'application canonique de l'espace transverse $U^{\vee}$ de $(T, T')_U$ dans l'espace transverse $E^{\vee}$ soit un homéomorphisme sur un ouvert de $E^{\vee}$; un tel voisinage sera appelé *ouvert distingué* de (T, T').

Le feuilletage (T, T') est dit *simple* si toute feuille est simple. Il est dit *localement simple* si tout point x de E admet un voisinage ouvert U relativement à T tel que $(T, T')_U$ soit simple; un tel ouvert U sera appelé *ouvert simple* et une feuille de $(T, T')_U$ sera appelée une *plaque* de U; alors tout point x admet un système fondamental de voisinages simples.

Soit (T, T') un feuilletage localement simple sur E. On appelle *chaîne simple* une suite finie $C = (U_1, \ldots U_n; V_{1,2}, \ldots, V_{n-1,n})$ vérifiant les conditions suivantes :

(1) Pour tout $i \leqslant n$, U_i est un ouvert simple de E.

(2) Pour tout $i < n$, $V_{i,i+1}$ est une plaque du feuilletage induit sur $U_i \cap U_{i+1}$ et $V_{i-1,i}$ et $V_{i,i+1}$ appartiennent à une même plaque V_i de U_i.

Il en résulte que la suite $(V_1, \ldots, V_n; V_{1,2}, \ldots, V_{n-1,n})$ est une chaîne de plaques d'une feuille F, c'est-à-dire que $V_{i,i+1}$ est une composante connexe de $V_i \cap V_{i+1}$. Si de plus x et x' sont deux points de F tels que $x \in V_1$ et $x' \in V_n$, on dira que C *relie* x à x'. Le triplet (x, C, x') est appelé *chaîne simple pointée.*

On appelle *chaîne pure* une suite $\Gamma = (W_1, \ldots, W_n; W_{1,2}, \ldots, W_{n-1,n})$ vérifiant les conditions suivantes :

(1) Pour tout $i \leqslant n$, W_i est un ouvert simple de (T, T').

(2) Pour tout $i < n$, $W_{i,i+1}$ est un ouvert distingué dans W_i et W_{i+1}.

(3) Pour tout $i < n$, W_i et W_{i+1} sont obtenus par saturation de $W_{i,i+1}$ relativement aux relations d'équivalence sous-jacentes à $(T, T')_{W_i}$ et à $(T, T')_{W_{i+1}}$ respectivement.

(4) Pour tout $i < n$, une plaque $V'_{i,i+1}$ de $W_{i,i+1}$ est une composante connexe de $V_i' \cap V_{i+1}'$, où V_i' est une plaque de W_i.

A Γ et à toute plaque V_1 de W_1 est associée une chaîne simple $(W_1, \ldots, W_n; V_{1,2}, \ldots, V_{n-1,n})$ bien déterminée, où $V_{1,2}$ est contenu dans V_1, et par suite une chaîne de plaques $(V_1, \ldots, V_n; V_{1,2}, \ldots, V_{n-1,n})$. La suite $(x, W_1, \ldots, W_n; W_{1,2}, \ldots, W_{n-1,n}, x') = (x, \Gamma, x')$, où $x \in V_1$ et $x' \in V_n$, est appelée *chaîne pure pointée* reliant x à x'.

PROPOSITION. *Avec les notations précédentes, l'application :* $V_1 \to V_n$ *est un isomorphisme de l'espace transverse* $W_1\check{}$ *sur l'espace transverse* $W_n\check{}$, *appelé isomorphisme canonique associé à* Γ.

PROPOSITION. *Si* (T, T') *est un feuilletage localement simple, la relation d'équivalence sous-jacente est ouverte.*

COROLLAIRE. *L'adhérence d'une feuille F relativement à la topologie T est une réunion de feuilles.*

Soit $(U_1, U_2, V_{1,2})$ une chaîne simple. Dans $U_1 \cap U_2$, il existe un ouvert W distingué relativement à U_1 et à U_2 et dont une plaque contient $V_{1,2}$. Soient W_1, W_2 et $W_{1,2}$ les ouverts obtenus par saturation de W relativement aux relations d'équivalence sous-jacentes à $(T, T')_{U_1}$, $(T, T')_{U_2}$, $(T, T')_{U_1 \cap U_2}$; alors $(W_1, W_2; W_{1,2})$ est une chaîne pure.

PROPOSITION. *A toute chaîne simple $(U_1, \ldots, U_n; V_{1,2}, \ldots, V_{n-1,n})$ est associée une chaîne pure $(W_1, \ldots, W_n; W_{1,2}, \ldots, W_{n-1,n})$ telle que $W_i \subset U_i$ et $V_{i,i+1} \subset W_{i,i+1}$.*

PROPOSITION. *Si (T, T') est un feuilletage topologique et si T' est régulière, toute feuille simple est propre. Si de plus (T, T') est localement simple, toute plaque d'un ouvert simple U est propre et les plaques forment une base de T'; pour qu'une feuille soit propre il faut et il suffit qu'il existe un ouvert simple dont la trace sur F soit connexe.*

PROPOSITION. *Pour qu'un feuilletage localement simple (T, T') soit simple, il faut et il suffit que pour tout $x \in E$ il existe un voisinage simple U de x relativement à T tel que deux plaques différentes de U appartiennent à deux feuilles différentes de (T, T').*

COROLLAIRE. *Si (T, T') est simple, tout feuilletage induit est simple et pour tout ouvert U de E relativement à T l'espace transverse $U^{\smile}$ de $(T, T')_U$ est un espace étalé (non séparé) au-dessus de $E^{\smile}$.*

La classe des feuilletages topologiques sur les ensembles, munie de la relation d'ordre : $(T_1, T_1') < (T, T')$ si, et seulement si, (T_1, T_1') est un feuilletage induit par (T, T'), est une espèce de structures locales au-dessus des ensembles (voir **3**).

Un feuilletage topologique peut être défini de la façon suivante : Soit E un ensemble muni d'une topologie T. La classe $\mathfrak{S}$ des sous-espaces (A, T_A), où T_A est la topologie induite par T sur le sous-ensemble A, forme une espèce de structures locales au-dessus des sous-ensembles de E pour la relation : $(A', T_{A'}) < (A, T_A)$ si, et seulement si, A' est ouvert pour T_A, et la projection $p : (A, T_A) \to A$. Désignons cette espèce de structures par $\langle \mathfrak{P}(E), p, \mathfrak{S} \rangle$.

$\langle \mathfrak{P}(E), p, \mathfrak{S} \rangle$ peut être complétée en une espèce de structures $\mathfrak{S}^-$ dont les éléments sont les classes complètes C de $\mathfrak{S}$ relativement à p, c'est-à-dire **(3)** les sous-classes de $\mathfrak{S}$ saturées par induction et agrégation et telles que, pour tout $(A, T_A) \in C$ et tout $(A', T_{A'}) \in C$, $A \cap A'$ soit ouvert relativement à T_A et à $T_{A'}$.

$\mathfrak{S}^-$ est une espèce de structures locales au-dessus de $\mathfrak{P}(E)$ pour la projection

$$\bar{p}: C \to \bigcup_{s \in C} p(s).$$

L'image $\bar{p}(C)$ de $C \in \mathfrak{S}^-$, munie de la topologie $\tau(C)$ engendrée par les éléments $p(s)$, où $s \in C$, est appelée *sous-espace local* de E. Le couple $(T_B, \tau(C))$, où T_B est la topologie induite par T sur $B = \bar{p}(C)$, définit un feuilletage topologique sur B. En particulier :

PROPOSITION. *La donnée d'un feuilletage topologique (T, T') sur E équivaut à la donnée d'un élément C de $\mathfrak{S}^-$ tel que $\bar{p}(C) = E$.*

Les ensembles $p(s)$, où $s \in C$, sont des ouverts de T'.

A l'espèce de structures locales $\langle \mathfrak{P}(E), p, \mathfrak{S} \rangle$ est associée la classe des jets locaux $j_x^\lambda s$, où $s \in \mathfrak{S}$ et $x \in p(s)$ (voir **5**). Le jet $j_x^\lambda s$ s'identifie à la classe des couples (s', x), où $s' \in \mathfrak{S}$ et $x \in p(s')$, tels que s et s' aient un ouvert commun contenant x. Ce jet sera appelé *germe de sous-espace en* x. Soit $E^\wedge$ l'ensemble de tous les germes de sous-espaces de E. Les ensembles $s^\wedge$ formés de tous les germes $j_x^\lambda s$ du sous-espace s forment une base d'une topologie sur $E^\wedge$; un élément $s^\wedge$ sera appelé *ouvert élémentaire.* L'application canonique $p^\wedge : j_x^\lambda s \to x$ est une application continue de $E^\wedge$ sur E dont la restriction à $s^\wedge$ est un homéomorphisme sur le sous-espace s.

Un ouvert W de $E^\wedge$ est appelé *espace extrait* de E (voir **4**). En particulier, pour que W soit l'ensemble des germes d'un sous-espace local B de E, il faut et il suffit que la restriction de $p^\wedge$ à W soit une application biunivoque sur B. Ainsi un feuilletage topologique (T, T') est aussi défini par la donnée d'un sous-espace extrait W de E tel que la restriction de $p^\wedge$ à W soit une application biunivoque sur E.

Il convient de considérer plus généralement les espaces extraits connexes relativement séparés W, c'est-à-dire tels que si $p^\wedge(w) = p^\wedge(w')$, où $w \in W$ et $w' \in W$, il existe deux voisinages de w et w' sans point commun.

Soit E' un espace topologique et f une application de E' dans l'espace topologique E. On dit que f est une *immersion* de E' dans E s'il existe un étalement f' de E' dans $E^{\wedge}$ tel que $f = p^{\wedge}f'$; c'est-à-dire pour tout $x \in E'$, il existe un voisinage ouvert U' de x dans E' tel que la restriction de f' à U' soit un homéomorphisme sur un ouvert de $E^{\wedge}$. Si de plus f est une application biunivoque, alors on dira que f est un *plongement* de E' dans E.

Soit E une variété topologique, de topologie T. Un sous-espace (A, T_A) de E est appelé une *sous-variété élémentaire* si T_A est une topologie de variété. Les sous-variétés élémentaires forment une sous-espèce de structures locales de $\mathfrak{S}$ et leurs germes, un sous-espace $E_v^{\wedge}$ de $E^{\wedge}$; $E_v^{\wedge}$ est une variété topologique (non séparée). Un ouvert de $E_v^{\wedge}$ sera appelé *variété extraite* de E.

Soient E et E' deux espaces topologiques et f une application de E' dans E. On appellera *point limite de E' relativement à f au-dessus de $x \in E$* un filtre minimal (c'est-à-dire le plus fin possible suivant l'ordre considéré sur la catégorie des filtres **(5)**) X de E' ayant les propriétés suivantes :

(1) $f(X)$ converge vers $x \in E$ et X ne converge pas vers un point y tel que $f(y) = x$.

(2) X admet une base B formée d'ouverts connexes.

(3) Pour tout $U' \in B$, il existe un ouvert U de E tel que, en désignant par $U^{\cdot\prime}$ la frontière de U', on ait : $U \cap f(U^{\cdot\prime}) = \emptyset$.

En particulier, on définit ainsi la notion de point limite d'un espace étalé; plus généralement, on appellera point limite d'un espace extrait W de E un point limite de W relativement à la restriction de $p^{\wedge}$ à W.

Soit (T, T') un feuilletage topologique sur E. On appellera *point limite d'une feuille* F de (T, T') un point limite X de l'espace topologique F relativement à l'injection canonique de F dans l'espace E muni de la topologie T. Alors X s'identifie à un point limite de l'espace extrait correspondant à F.

PROPOSITION. *Soient (T, T') un feuilletage localement simple et U un ouvert simple. Pour qu'une plaque P de U n'admette pas de point limite, il faut et il suffit que P soit fermée dans U pour la topologie induite par T sur U.*

PROPOSITION. *Soit (T, T') un feuilletage localement simple. Pour* *qu'une feuille F n'admette pas de point limite, il faut et il suffit que, pour tout ouvert simple U, toute plaque de $F \cap U$ soit fermée dans U.*

Si aucune feuille de (T, T') n'admet de point limite, on dira que (T, T') est un *feuilletage sans point limite.*

Si une feuille F d'un feuilletage (T, T') admet un point limite, l'espace transverse de E n'est pas séparé. Toutefois un feuilletage sans point limite peut être tel que son espace transverse ne soit pas localement séparé.

2. *Partitions locales*

Soit E un ensemble muni d'une topologie T. Soit U un ouvert de E et soit $\rho(U)$ une relation d'équivalence ouverte sur U telle que les classes d'équivalence modulo $\rho(U)$ soient connexes, localement connexes et fermées pour la topologie T_U induite par T sur U. La classe de toutes les relations d'équivalence $\rho(U)$, où U parcourt l'ensemble des ouverts de E, est une classe inductive $\mathfrak{R}$ pour la relation d'ordre : $\rho(U') < \rho(U)$ si, et seulement si, $U' \subset U$ et si les classes d'équivalence modulo $\rho(U')$ sont les composantes connexes des traces sur U' des classes d'équivalence modulo $\rho(U)$. Toute sous-classe majorée $\mathfrak{R}'$ de $\mathfrak{R}$ admet, en effet, un agrégat, à savoir la relation d'équivalence engendrée par les relations $\rho(U) \in \mathfrak{R}'$. L'application $\mathrm{p} : \rho(U) \to U$ étale $\mathfrak{R}$ sur la classe inductive des ouverts de E.

Une classe complète C de $\mathfrak{R}$, compatible avec p et telle que

$$\bigcup_{s \in C} p(s) = E$$

sera appelée *partition locale* de E. Une classe d'équivalence modulo s, où $s \in C$, sera appelée *plaque* de C.

PROPOSITION. *Soit C une partition locale de E et ρ la relation d'équivalence sur E engendrée par les relations d'équivalence $s \in C$. Alors les plaques de C forment une base d'une topologie T' sur E telle que (T, T') soit un feuilletage topologique sans point limite dont ρ est la relation d'équivalence sous-jacente.*

Le feuilletage (T, T') défini par la proposition sera appelé *feuilletage associé à la partition locale C.*

Les partitions locales forment une espèce de structures locales au-dessus des espaces topologiques, la relation d'ordre étant la relation d'inclusion.

Soient (T, T') un feuilletage topologique sur E et ρ la relation d'équivalence sous-jacente. Les composantes connexes des ouverts relativement à T des feuilles de (T, T') forment une base d'une topologie T'' sur E.

PROPOSITION. *Avec les notations précédentes, T'' est une topologie plus fine que T et moins fine que T'. Si (T, T') est un feuilletage localement simple sans point limite, T' et T'' sont identiques et (T, T') est le feuilletage associé à une partition locale.*

De cette proposition, il résulte que (T'', T') est un feuilletage topologique. Si (T, T') est localement simple et sans point limite, c'est le feuilletage associé à la partition locale de E dont une base est formée des relations d'équivalence sous-jacentes à $(T, T')_U$, où U est un ouvert simple de E.

Soit C' une partition locale de l'espace topologique E admettant une base C telle que pour tout $x \in p(s) \cap p(s')$, où $s \in C$ et $s' \in C$, il existe un ouvert $U \ni x$ sur lequel les traces des classes modulo s et s' respectivement soient connexes, c'est-à-dire identiques aux classes d'équivalence modulo $s'' \in C$, où $p(s'') = U$.

Alors C s'identifie à un *système dynamique généralisé* au sens de Reeb (voir **14**).

PROPOSITION. *Pour qu'une partition locale soit un système dynamique généralisé, il faut et il suffit que le feuilletage associé soit localement simple.*

Les notions de système dynamique généralisé et de feuilletage topologique localement simple sans point limite sont donc équivalentes.

Soient (T, T') un feuilletage topologique sur E et F une feuille localement compacte. On appellera *point adhérent à l'infini de F* un point adhérent relativement à T au filtre d'Alexandroff de F (c'est-à-dire au filtre admettant pour base les complémentaires des compacts de F relativement à la topologie T'). Plus précisement, on peut considérer les points de F adhérents, relativement à T, à un filtre $\mathfrak{F}$ de F tel que :

(1) $\mathfrak{F}$ est plus fin que le filtre d'Alexandroff.

(2) $\mathfrak{F}$ admet une base formée d'ouverts à frontière compacte.

Un point adhérent à un tel filtre $\mathfrak{F}$ est un point limite de F selon $\mathfrak{F}$ au sens de Reeb **(14)**.

PROPOSITION. *Soient* (T, T') *un feuilletage localement simple sans point limite,* F *une feuille localement compacte et* $\mathfrak{F}$ *un filtre sur* F *vérifiant les conditions* (1) *et* (2) *ci-dessus. Alors l'ensemble des points adhérents à* F *relativement à* $\mathfrak{F}$ *est une réunion de feuilles.*

Pour la démonstration, voir **(14)**.

En particulier, cette proposition s'applique si $\mathfrak{F}$ est un bout au sens de Freudenthal **(15)**.

Remarquons qu'une feuille F peut être contenue dans son adhérence à l'infini (la feuille est alors dite stable au sens de Poisson).

3. *Variétés feuilletées*

Soient G un pseudogroupe de transformations dans l'espace numérique R^p et G' un pseudogroupe de transformations dans l'espace numérique R^q. Soient $\tau(G)$ et $\tau(G')$ les topologies sous-jacentes à G et à G' respectivement (voir **3**). Soit h un homéomorphisme $(x, y) \to (g(x), g_x'(y))$, défini sur $U \times U'$, où U est la source de $g \in G$ et U' la source de $g_x' \in G'$ pour tout $x \in U$, tel que de plus $h(U \times U')$ soit ouvert pour la topologie produit $\tau(G) \times \tau(G')$. L'ensemble des homéomorphismes h est la base d'un pseudogroupe de transformations $G \# G'$ dans $R^p \times R^q$. Tout point de R^p est supposé fermé pour $\tau(G)$.

Soit A un atlas complet de $R^p \times R^q$ sur un ensemble E compatible avec le pseudogroupe $G \# G'$ (voir **6**). On dira que l'atlas complet A définit sur E une *structure feuilletée de l'espèce* $G \# G'$ *élargie*, c'est-à-dire obtenue par élargissement complet de $G \# G'$ au-dessus du pseudogroupe $\mathfrak{E}$ des applications biunivoques d'un ensemble sur un ensemble. Considérons en particulier l'ensemble A_1 des cartes locales $f \in A$ ayant pour source un ouvert élémentaire $U \times U'$, où $U \in \tau(G)$ et $U' \in \tau(G')$; une telle carte sera appelée *carte élémentaire*. A_1 est une base de l'atlas complet A. Soit $\tau(A)$ la topologie sur E sous-jacente à A, c'est-à-dire la topologie dont les buts des cartes de A forment une base; $\tau(A)$ admet aussi pour base l'ensemble des buts des cartes élémentaires.

A une carte élémentaire f définie sur $U \times U'$ correspond, pour tout $x \in U$, l'application $f_x : y \rightarrow f(x, y)$ de U' dans E. L'ensemble des applications f_x forme une base d'un atlas complet A' de R^q dans E compatible avec G'; cet atlas complet A' définit sur E une structure de l'espèce G' élargie. Soit $\tau(A')$ la topologie sous-jacente à A'; nous supposerons $\tau(G')$, donc $\tau(A')$, localement connexe.

PROPOSITION. *Avec les notations précédentes, le couple des topologies* $(\tau(A), \tau(A'))$ *définit sur E un feuilletage topologique localement simple sans point limite, sous-jacent à la structure feuilletée de l'espèce* $G \# G'$ *élargie.*

Les composantes connexes de $\tau(A')$ sont aussi appelées les feuilles de la structure feuilletée de l'espèce $G \# G'$ élargie. Chaque feuille est munie d'une structure de l'espèce G' élargie.

En particulier, si G et G' sont les pseudogroupes de tous les automorphismes locaux de R^p et R^q respectivement, une structure feuilletée de l'espèce $G \# G'$ élargie sera appelée structure de *variété feuilletée.* Le psuedogroupe $G \# G'$ correspondant sera désormais noté $\Lambda_{p,q}$.

PROPOSITION. *Si E est une variété feuilletée telle que le feuilletage topologique sous-jacent soit simple, l'espace transverse* $E^\check{}$ *est une variété non séparée.*

Soit A un atlas complet définissant sur E une structure feuilletée de l'espèce $G \# G'$ élargie. A est aussi compatible avec $\Lambda_{p,q}$. En complétant A relativement à $\Lambda_{p,q}$, on obtient un atlas complet $A_{p,q}$ définissant sur E une structure de variété feuilletée sous-jacente à la structure de l'espèce $G \# G'$ élargie. Chaque feuille F est aussi munie d'une structure de variété de dimension q par l'atlas $A_{p,q}'$ correspondant à $A_{p,q}$ par la même construction que précédemment. De plus, F est une réunion de feuilles de la variété feuilletée sous-jacente.

Dans tout ce qui précède, on peut remplacer $G \# G'$ par un sous-pseudogroupe Γ et définir une structure feuilletée de l'espèce Γ élargie par la donnée d'un atlas complet compatible avec Γ. Une structure feuilletée de l'espèce Γ élargie admet une structure feuilletée sous-jacente de l'espèce $G \# G'$ élargie.

Exemples de structures feuilletées de l'espèce Γ élargie

(1) Si Γ est le pseudogroupe admettant pour base les homéomorphismes $h : (x, y) \to (g(x), g'(y))$ de $U \times U'$ sur un ouvert de $R^p \times R^q$, où U est la source de $g \in G$ et U' la source de $g' \in G'$, une structure feuilletée de l'espèce Γ élargie est appelée structure de *produit local.* Une telle structure admet deux structures feuilletées sous-jacentes, de l'espèce $G \# G'$ et de l'espèce $G' \# G$ élargies. On dira aussi qu'une structure de produit local définit un *feuilletage double.*

(2) Soit $\Lambda_{p,q}{}^{k,l,r}$ le sous-pseudogroupe de $\Lambda_{p,q}$ admettant pour base les homéomorphismes $h : (x, y) \to (g(x), g_x'(y))$ de $\Lambda_{p,q}$ tels que $g \in \Lambda_p{}^k$, $g_x' \in \Lambda_q{}^l$ pour tout x appartenant à la source U de g et $h \in \Lambda_{p+q}{}^r$, où $r < k$, $r < l$ et $\Lambda_n{}^s$ désigne le pseudogroupe des automorphismes locaux d'ordre s de l'espace numérique R^n. Les structures feuilletées de l'espèce $\Lambda_{p,q}{}^{k,l,r}$ élargie sont appelées *structures feuilletées différentiables d'ordre* (k, l, r). Les feuilles sont munies de structures de variétés k fois différentiables.

Toute structure de variété feuilletée différentiable d'ordre (r, r, r) telle que le feuilletage topologique sous-jacent soit simple, est isomorphe à une structure de variété feuilletée construit de la façon suivante:

Soient V_n et B deux variétés r fois différentiables, B pouvant ne pas être séparée. Soit f une application de V_n sur B de rang p constant. Soient $x_0 \in V$ et $y_0 = f(x_0)$. Alors il existe une carte différentiable distinguée g du pavé ouvert de R^n formé des points $(x_1, \ldots, x_n)$ tels que $a_i < x_i < b_i$ pour tout $i \leqslant n$ sur un voisinage ouvert U de x_0 dans V_n et une carte différentiable g' du pavé de R^q formé des points $(y_1, \ldots, y_q)$ tels que $a_j' < y_j < b_j'$ pour tout $j \leqslant k$ sur un voisinage ouvert V de y_0 dans B de façon que l'application $f' = g'^{-1}fg$ soit définie par :

$$y_i = x_i \text{ si } i \leqslant p \quad \text{et} \quad y_j = 0 \text{ si } p < j < q.$$

Pour deux cartes distinguées g et g_1 de R^n dans V_n le changement de cartes $g^{-1}g_1$ appartient à $\Lambda_{p,n-p}{}^{r,r}$. Donc l'ensemble des cartes distinguées de R^n dans V_n définit sur V_n une structure de variété feuilletée différentiable d'ordre (r, r, r).

Les feuilles sont les composantes connexes des ensembles $f^{-1}(y)$, où $y \in B$; elles forment une partition de V_n et l'espace quotient de V_n par cette partition est une variété non séparée de dimension p se projetant sur une variété extraite de dimension p de B.

(3) Si, dans l'exemple précédent, on remplace $\Lambda_p{}^k$, $\Lambda_q{}^l$ et $\Lambda_{p+q}{}^r$ par les pseudogroupes d'automorphismes locaux analytiques réels (ou complexes) $\Lambda_p{}^\omega$, $\Lambda_q{}^\omega$, $\Lambda_{p+q}{}^\omega$ de R^p, R^q et R^{p+q} respectivement les structures feuilletées de l'espèce $\Lambda_{p,q}{}^\omega$ élargie correspondantes sont appelées *structures feuilletées analytiques* réelles (ou complexes). Toute structure feuilletée analytique admet une structure feuilletée sous-jacente, différentiable d'ordre (k, l, r) pour tout k, l, r.

Si $\Lambda_q{}^l$ (respectivement $\Lambda_p{}^k$) est remplacé par $\Lambda_q{}^\omega$ (respectivement par $\Lambda_p{}^\omega$), on obtient une structure feuilletée à feuilles analytiques (respectivement transversalement analytique).

4. *Structures feuilletées de l'espèce $\mathfrak{B} \# \mathfrak{F}$ élargie*

Soit $\mathfrak{E}^\sim$ la catégorie inductive des applications d'un ensemble dans un autre, un élément étant représenté par le triplet (E', f, E), où f est une application de E dans E'. Le groupoïde des éléments inversibles est le groupoïde $\mathfrak{E}$ des applications biunivoques d'un ensemble sur un autre.

Soient $\mathfrak{B}$ et $\mathfrak{F}$ deux groupoïdes locaux au-dessus de $\mathfrak{E}$ (voir **5**), les foncteurs* projections étant p et p' respectivement, c'est-à-dire $\langle \mathfrak{E}, p, \mathfrak{B} \rangle$ et $\langle \mathfrak{E}, p', \mathfrak{F} \rangle$ sont des espèces de structures locales au-dessus de $\mathfrak{E}$. Les symboles S et s désigneront des unités (ou structures) de $\mathfrak{B}$, les symboles S' et s' des unités de $\mathfrak{F}$; nous écrirons aussi $B = p(S)$ et $F = p'(S')$; un élément de $\mathfrak{B}$ sera représenté par un couple (g, S), où $g \in \mathfrak{E}$ et $\alpha(g) = p(S)$; un élément de $\mathfrak{F}$ par un couple (g', S'), où $g' \in \mathfrak{E}$ et $\alpha(g') = p'(S')$. Les éléments $p(s)$, où $s < S$, engendrent une topologie $\tau(S)$ sur B et les éléments $p'(s')$, ou $s' < S'$, une topologie $\tau(S')$ sur F, supposée localement connexe.

Soit $\widetilde{\mathfrak{B} \# \mathfrak{F}}$ la catégorie des triplets $(S_2 \times S_2', h, S_1 \times S_1')$, où h est un homéomorphisme de $B_1 \times F_1$ sur un ouvert de $B_2 \times F_2$ pour les topologies $\tau(S_1) \times \tau(S_1')$ et $\tau(S_2) \times \tau(S_2')$, de la forme :

$$(x, y) \rightarrow (g(x), g_x'(y))$$

*Le mot foncteur signifie toujours foncteur covariant. La classe des unités ou des objets d'une catégorie $\mathfrak{C}$ sera désignée par $\mathfrak{C}_0$, l'unité à droite, ou source, de $f \in \mathfrak{C}$ par $\alpha(\mathfrak{f})$, l'unité à gauche, ou but, de f par $\beta(f)$.

où $(g, S_1) \in \mathfrak{B}$, $(x, y) \in B_1 \times F_1$, $(g_x', S_1') \in \mathfrak{F}$ pour tout $x \in \mathfrak{B}$, avec $\beta(g, S_1) < S_2$ et $\beta(g_x', S_1') < S_2'$. Munie de la relation : $(s_2 \times s_2', h', s_1 \times s_1') < (S_2 \times S_2', h, S_1 \times S_1')$ si, et seulement si, $s_i < S_i$, $s_i' < S_i'$ où $i = 1, 2$, et h' est une restriction de h, $\widetilde{\mathfrak{B} \# \mathfrak{F}}$ devient une catégorie sous-inductive presque au-dessus de $\mathfrak{E}^\sim$ (voir **3**). Soit $\mathfrak{B} \# \mathfrak{F}$ le groupoïde des éléments inversibles.

Soient E un ensemble et f une application biunivoque de $B \times F$ sur E. Soit $(f, S \times S')\mathfrak{B} \# \mathfrak{F}$ la classe des couples $(fh, S_1 \times S_1')$, où $(S \times S', h, S_1 \times S_1') \in \mathfrak{B} \# \mathfrak{F}$. Une telle classe sera appelée *structure de produit assoupli sur E.*

Soit $\sigma = (f, S \times S')\mathfrak{B} \# \mathfrak{F}$ une structure de produit assoupli sur $\pi(\sigma) = E$; on dira que $\sigma' = (f', s \times s')\mathfrak{B} \# \mathfrak{F}$ est *induite* par σ si, et seulement si, $\pi(\sigma') \subset \pi(\sigma)$ et s'il existe un élément $(S \times S', h, s \times s')$ de $\mathfrak{B} \# \mathfrak{F}$ tel que $f' = fh$. Avec cette relation, la classe de toutes les structures de produit assoupli devient une classe ordonnée (**3**).

Deux structures de produits assouplis σ et σ' sont dites *localement compatibles* si la classe Σ'' de toutes les structures de produits assouplies σ'' induites simultanément par σ et σ' est telle que l'on ait :

$$\pi(\sigma) \cap \pi(\sigma') = \bigcup_{\sigma'' \in \Sigma''} \pi(\sigma'').$$

Soit Σ une classe complète de produits assouplis σ, c'est-à-dire une classe saturée par induction et agrégation et telle que deux quelconques de ses éléments soient localement compatibles. Alors Σ est appelée *structure feuilletée de l'espèce* $\mathfrak{B} \# \mathfrak{F}$ *élargie sur*

$$E = \bigcup_{\sigma \in \Sigma} \pi(\sigma).$$

A une structure de produit assoupli σ correspond la structure feuilletée $\bar{\sigma}$ définie par la classe complète engendrée par la classe de toutes les structures de produits assouplis qu'elle induit. Remarquons que $\bar{\sigma}$ peut contenir d'autres structures de produits assouplis sur $\pi(\sigma)$, qui admettent par conséquent la même structure feuilletée sous-jacente.

La classe des structures feuilletées de l'espèce $\mathfrak{B} \# \mathfrak{F}$ élargie, munie de la relation : $\Sigma < \Sigma'$ si, et seulement si, la classe Σ est contenue dans Σ', et de la projection

$$\bar{\pi}: \Sigma \to \bigcup_{\sigma \in \Sigma} \pi(\sigma) = E,$$

est une espèce de structures locales $\mathfrak{L}(\mathfrak{B} \# \mathfrak{F})_0$ au-dessus de $\mathfrak{E}$.

A tout $\Sigma \in \mathfrak{L}(\mathfrak{B} \# \mathfrak{F})_0$ correspond une topologie sous-jacente $\tau(\Sigma)$ ayant pour base les ensembles $\pi(\sigma)$, où $\sigma \in \Sigma$; alors Σ induit une structure feuilletée de l'espèce $\mathfrak{B} \# \mathfrak{F}$ élargie sur tout ouvert de cette topologie.

Soient $\sigma = (f, s \times s')\mathfrak{B} \# \mathfrak{F}$ et σ' deux produits assouplis. Un *isomorphisme* de σ sur σ' est un couple (k, σ), où $k \in \mathfrak{E}$, $\alpha(k) = \pi(\sigma)$ et $\sigma' = (kf, s \times s')\mathfrak{B} \# \mathfrak{F}$. Le groupoïde de ces isomorphismes est un groupoïde ordonné pour la relation : $(k_1, \sigma_1) < (k, \sigma)$ si, et seulement si, $\sigma_1 < \sigma$ et k_1 est une restriction de k. Ce groupoïde est l'élargissement de $\mathfrak{B} \# \mathfrak{F}$ relativement à $\mathfrak{E}$ **(3)**.

Soient $\Sigma \in \mathfrak{L}(\mathfrak{B} \# \mathfrak{F})_0$ et $\Sigma' \in \mathfrak{L}(\mathfrak{B} \# \mathfrak{F})_0$. On appelle *isomorphisme* de Σ sur Σ' un couple (K, Σ) tel que K soit une application biunivoque de $\bar{\pi}(\Sigma)$ sur $\bar{\pi}(\Sigma')$ et que, pour tout $\sigma \in \Sigma$, on ait $\beta(K\pi(\sigma), \sigma) \in \Sigma'$, où $K\pi(\sigma)$ est la restriction de K à $\pi(\sigma)$. Les isomorphismes ainsi définis forment un groupoïde local $\mathfrak{L}(\mathfrak{B} \# \mathfrak{F})$ au-dessus de $\mathfrak{E}$.

Soit $\Sigma \in \mathfrak{L}(\mathfrak{B} \# \mathfrak{F})_0$ et $(f, S \times S')\mathfrak{B} \# \mathfrak{F} \in \Sigma$; soit $x \in B = p(S)$. La classe ψ des éléments (f_x, S'), où $f_x(y) = f(x, y)$, est base d'un atlas complet A sur $E = \bar{\pi}(\Sigma)$, dont la topologie sous-jacente $\tau(A)$ est plus fine que $\tau(\Sigma)$. Le couple $(\tau(\Sigma), \tau(A))$ définit sur E un feuilletage topologique sous-jacent à Σ et A définit sur E une structure de l'espèce $\mathfrak{F}$ élargie. Chaque feuille de $(\tau(\Sigma), \tau(A))$, où $\tau(A)$ est localement connexe, est aussi munie d'une structure de l'espèce $\mathfrak{F}$ élargie.

PROPOSITION. *Soient $\mathfrak{B}'$ et $\mathfrak{F}'$ deux groupoïdes locaux au-dessus de $\mathfrak{B}$ et $\mathfrak{F}$ respectivement; alors $\mathfrak{L}(\mathfrak{B}' \# \mathfrak{F}')$ est un groupoïde local au-dessus de $\mathfrak{L}(\mathfrak{B} \# \mathfrak{F})$.*

Si $\mathfrak{B}$ et $\mathfrak{F}$ sont les sous-groupoïdes des éléments inversibles des catégories locales $\mathfrak{B}^{\sim}$ et $\mathfrak{F}^{\sim}$ au-dessus de $\mathfrak{E}^{\sim}$, une construction générale analogue mène à une catégorie locale $\mathfrak{L}(\mathfrak{B}^{\sim} \# \mathfrak{F}^{\sim})$, contenant $\mathfrak{L}(\mathfrak{B} \# \mathfrak{F})$ et dont les éléments sont des homomorphismes de structures feuilletées de l'espèce $\mathfrak{B} \# \mathfrak{F}$ élargie.

Toutes les constructions précédentes peuvent se faire de même en remplaçant $\widetilde{\mathfrak{B} \# \mathfrak{F}}$ et $\mathfrak{B} \# \mathfrak{F}$ par $\Gamma^{\sim}$ et Γ, où $\Gamma^{\sim}$ est une sous-catégorie de $\widetilde{\mathfrak{B} \# \mathfrak{F}}$ ayant les mêmes unités et saturée par induction, et Γ est le groupoïde des éléments inversibles de $\Gamma^{\sim}$. On est ainsi conduit à l'espèce $\mathfrak{L}(\Gamma)_0$ des structures feuilletées de l'espèce Γ élargie. Toute structure feuilletée de l'espèce Γ élargie admet une structure feuilletée sous-jacente de l'espèce $\mathfrak{B} \# \mathfrak{F}$ élargie.

Exemples de structures feuilletées de l'espèce Γ *élargie*

(1) *Produits locaux.* Si Γ est le sous-pseudogroupe $\mathfrak{B} \times \mathfrak{F}$ de $\mathfrak{B} \# \mathfrak{F}$, une structure feuilletée de l'espèce $\mathfrak{B} \times \mathfrak{F}$ élargie est appelée structure de *produit local.* L'espèce $\mathfrak{L}(\mathfrak{B} \times \mathfrak{F})_0$ admet $\mathfrak{L}(\mathfrak{B} \# \mathfrak{F})_0$ et $\mathfrak{L}(\mathfrak{F} \# \mathfrak{B})_0$ comme espèces de structures locales sous-jacentes. Les structures feuilletées de l'espèce $\mathfrak{B} \# \mathfrak{F}$ élargie et de l'espèce $\mathfrak{F} \# \mathfrak{B}$ élargie sous-jacentes à une structure de produit local seront dites *supplémentaires.*

(2) *Fibrations.* Si on suppose que $\mathfrak{F}$ est un groupoïde muni de la relation d'ordre triviale ($a < b$ si, et seulement si, $a = b$ ou $a = 0$), la topologie $\tau(S')$ est la topologie grossière sur $F = p'(S')$. Dans ce cas, une structure feuilletée de l'espèce $\mathfrak{B} \# \mathfrak{F}$ élargie est appelée *structure fibrée* et les feuilles sont appelées les fibres. En particulier, supposons que $\langle \mathfrak{T}, p'', \mathfrak{F} \rangle$ soit une espèce de superstructures locales au-dessus de $\langle \mathfrak{E}, \tau, \mathfrak{T} \rangle$, où $\mathfrak{T}$ est le gruopoïde local des homéomorphismes d'un espace topologique sur un autre; c'est-à-dire l'ensemble F est muni d'une topologie $p''(S')$ et l'application $p'' : (g', S') \to (g', p''(S'))$ est un foncteur de $\mathfrak{F}$ vers $\mathfrak{T}$. Toute structure fibrée Σ sur E admet pour structure sous-jacente une structure feuilletée de l'espèce $\mathfrak{B} \# \mathfrak{T}$ élargie. Si les topologies $p''(S')$ sont localement connexes, les feuilles de cette structure feuilletée sont les composantes connexes des fibres.

Pour tout sous-pseudogroupe Γ de $\mathfrak{B} \# \mathfrak{F}$, on a aussi les structures fibrées de l'espèce Γ élargie.

Soit $\mathfrak{B}$ un pseudogroupe d'automorphismes locaux de l'espace topologique S, l'ensemble sous-jacent à S étant B; soit $\mathfrak{F}$ un groupe topologique opérant d'une façon continue sur un espace topologique S', l'ensemble sous-jacent à S' étant F. Soit Γ le sous-pseudogroupe de $\mathfrak{B} \# \mathfrak{F}$ formé des éléments $(S_2 \times S', h, S_1 \times S')$ de $\mathfrak{B} \# \mathfrak{F}$ tels que l'application $h : (x, y) \to (g(x), g_x'(y))$ vérifie de plus la condition : $x \to g_x'$ est continu. Une structure fibrée de l'espèce Γ élargie est alors une structure fibrée à *groupe structural topologique.*

On définit de même les structures fibrées différentiables, analytiques, etc.

Variation de structures. Soient $\mathfrak{B}'$ et $\mathfrak{F}'$ des groupoïdes locaux au-dessus de $\mathfrak{B}$ et $\mathfrak{F}$ respectivement et π' la projection correspondante de $\mathfrak{B}' \# \mathfrak{F}'$ vers $\mathfrak{B} \# \mathfrak{F}$. Soient Γ et Γ' deux sous-groupoïdes de $\mathfrak{B} \# \mathfrak{F}$ et $\mathfrak{B}' \# \mathfrak{F}'$ respectivement tels que $\pi'(\Gamma') \subset \Gamma$, les sous-groupoïdes Γ et Γ' définissant les espèces de structures feuilletées $\mathfrak{L}(\Gamma)_0$ et $\mathfrak{L}(\Gamma')_0$. Alors $\mathfrak{L}(\Gamma')$ est un groupoïde local au-dessus de $\mathfrak{L}(\Gamma)$ relativement à une projection η. On est conduit au problème suivant : étant donné $\Sigma \in \mathfrak{L}(\Gamma)_0$, étudier les structures $\Sigma^- \in \mathfrak{L}(\Gamma')_0$ telles que $\Sigma = \eta(\Sigma^-)$.

Plus précisément : supposons $\mathfrak{F}'$ étalé (voir **3**) au-dessus de $\mathfrak{F}$ et $\mathfrak{B}' = \mathfrak{B}$; on a alors le problème d'existence et d'unicité à un isomorphisme près d'une structure Σ^- telle que $\Sigma = \eta(\Sigma^-)$ et qu'une feuille donnée F' de Σ soit munie par Σ^- d'une structure donnée $\bar{S}'$ de l'espèce $\mathfrak{F}_0'$. En particulier, si Σ est un produit assoupli, sous quelles conditions Σ^-, s'il existe, est-il un produit assoupli (cas de rigidité locale ou globale de la structure $\bar{S}'$) ? Ce problème est le problème de variation de structures relativement à une espèce de structures feuilletées sous-jacentes. Le problème de variations de structures complexes étudié par Kodaïra et Spencer correspond au cas suivant : $\mathfrak{L}(\Gamma)_0$ est formé des structures de variétés feuilletées différentiables (se réduisant en particulier à des espaces fibrés différentiables) et $\mathfrak{L}(\Gamma')_0$ est formé des structures de variétés feuilletées à feuilles analytiques complexes, les structures transversales étant différentiables.

5. *Feuilletages de seconde espèce*

Soient $\mathfrak{B}$ un groupoïde inductif complet au-dessus de $\mathfrak{T}$ et $\mathfrak{Z}$ un groupoïde inductif complet étalé au-dessus de $\mathfrak{T}$, les projections étant p et p_1 respectivement. Soit Θ la classe des triplets (S, f, Z) tels que $S \in \mathfrak{B}_0$, $Z \in \mathfrak{Z}_0$, $(p(S), f, p_1(Z)) \in \mathfrak{T}^\sim$ = catégorie locale des applications continues. Posons :

$$\alpha(S, f, Z) = Z,\ \beta(S, f, Z) = S,\ q(S, f, Z) = (p(S), f, p_1(Z)).$$

Θ est muni d'une relation d'ordre : $(S', f', Z) < (S, f, Z)$ si, et seulement si, $Z' < Z$ dans $\mathfrak{Z}$, $S' < S$ dans $\mathfrak{B}$, $q(S', f', Z') < q(S, f, Z)$ dans $\mathfrak{T}^\sim$. $\mathfrak{B}$ et $\mathfrak{Z}$ opèrent sur Θ pour les lois de composition:

$$[(k, S), (S, f, Z)] \to (S', kf, Z) \quad \text{où} \quad S' = \beta(k, S),\ (k, S) \in \mathfrak{B},$$
$$[(g, Z), (S, f, Z)] \to (S, fg^{-1}, Z') \quad \text{où} \quad Z' = \beta(g, Z),\ (g, Z) \in \mathfrak{Z}.$$

Θ est une classe locale au-dessus de $\mathfrak{T}^{\sim}$ relativement à la projection q.

Soit $\mathfrak{H}$ une sous-classe de Θ, munie de l'ordre induit et vérifiant les conditions suivantes :

(1) Si $h \in \mathfrak{H}$ tout composé de h avec un élément de $\mathfrak{B}$ ou de $\mathfrak{Z}$ appartient à $\mathfrak{H}$;

(2) Pour tout $z < Z$, soit H' la classe des $h' \in \mathfrak{H}$ tels que $h' < (S, f, Z) \in \mathfrak{H}$ et $\alpha(q(h')) < p_1(z)$; alors $p_1(z) = \cup\alpha(q(H'))$;

(3) Pour tout $s < S$, soit H'' la classe des $h'' \in \mathfrak{H}$ tels que $h'' < (S, f, Z) \in \mathfrak{H}$ et $\beta(q(h'')) < p(s)$; alors $p(s) = \cup\beta(q(H''))$.

Remarque. Si $\mathfrak{H}$ ne vérifie pas 1, on obtient par élargissement relativement à $\mathfrak{B}$ et $\mathfrak{Z}$ une classe vérifiant 1, 2, 3. De plus, si $\mathfrak{B}$ et $\mathfrak{Z}$ ne sont pas complets au-dessus de $\mathfrak{T}$, on peut les remplacer par leurs complétés relativement à $\mathfrak{T}$ (voir **3**).

Soit C une classe complète de $\mathfrak{H}$, localement compatible avec q, c'est-à-dire une sous-classe de $\mathfrak{H}$ saturée par induction et agrégation et telle que: si $h \in C$ et $h' \in C$, $q(h) \cap q(h') = \cup q(H')$, où H' est la classe des $h'' \in C'$, $h'' < h$ et $h'' < h'$; $\cup\, q(C)$ existe. C sera identifiée au triplet $(S, f, Z) \in \Theta$, où $Z = \cup\alpha(C)$, $S = \cup\beta(C)$ et

$$(p(S), f, p_1(Z)) = \cup q(C),$$

et appelée *homomorphisme* de Z vers S.

La classe $\mathfrak{H}^-$ des homomorphismes de $Z \in \mathfrak{Z}_0$ vers $S \in \mathfrak{B}_0$ munie de l'ordre défini par l'inclusion s'identifie à une sous-classe locale de Θ sur laquelle $\mathfrak{B}$ et $\mathfrak{Z}$ opèrent aussi. $\mathfrak{H}^-$ est une classe locale au-dessus de $\mathfrak{T}^{\sim}$ pour la projection q; de plus $\mathfrak{H}^-$ est étalé sur $\mathfrak{Z}_0$ et $\mathfrak{B}_0$ par les projections α et β.

Soit $\mathfrak{H}^{\sim}$ la réunion de $\mathfrak{H}^-$, de $\mathfrak{B}$ et de $\mathfrak{Z}$ (en supposant $\mathfrak{B}$ et $\mathfrak{Z}$ disjoints; sinon on prend la classe somme de $\mathfrak{H}^-$, de $\mathfrak{B}$ et de $\mathfrak{Z}$). Munissons $\mathfrak{H}^-$ de la loi de composition suivante : $h'h$ est défini si, et seulement si, l'une des conditions suivantes est vérifiée :

(1) $h' \in \mathfrak{Z}$, $h \in \mathfrak{Z}$; alors $h'h$ est le composé dans $\mathfrak{Z}$;

(2) $h' \in \mathfrak{B}$, $h \in \mathfrak{B}$; alors $h'h$ est le composé dans $\mathfrak{B}$;

(3) $h' = (S, f, Z) \in \mathfrak{H}^-$ et $h^{-1} = (g, Z) \in \mathfrak{Z}$; alors $h'h = (S, fg^{-1}, Z')$ où $Z' = \beta(g, Z)$;

(4) $h' = (k, S) \in \mathfrak{B}$ et $h = (S, f, Z) \in \mathfrak{H}^-$; alors $h'h = (S', kf, Z)$, où $S' = \beta(k, S)$.

$\mathfrak{H}^{\sim}$ devient ainsi une catégorie au-dessus de $\mathfrak{T}^{\sim}$ pour le foncteur projection $q^{\sim}$ dont la restriction à $\mathfrak{B}$ est p, la restriction à $\mathfrak{Z}$ est p_1

et la restriction à $\mathfrak{H}^-$ est q. De plus $\mathfrak{H}^\sim$ est une catégorie locale au-dessus de $\mathfrak{T}^\sim$ pour le foncteur $q^\sim$ si on munit $\mathfrak{H}^\sim$ de la loi d'induction suivante : $h < h'$ si, et seulement si, l'une des conditions suivantes est verifiée :

(1) $h \in \mathfrak{Z}$, $h' \in \mathfrak{Z}$ et $h < h'$ dans $\mathfrak{Z}$;

(2) $h \in \mathfrak{B}$, $h' \in \mathfrak{B}$ et $h < h'$ dans $\mathfrak{B}$;

(3) $h \in \mathfrak{H}^-$, $h' \in \mathfrak{H}^-$ et $h < h'$ dans $\mathfrak{H}^-$.

$\mathfrak{B}$ et $\mathfrak{Z}$ sont des sous-groupoïdes de $\mathfrak{H}^\sim$ saturés par induction.

Soit ρ la relation d'équivalence sur $\mathfrak{H}^-$: $(S, f, Z) \sim (S', f', Z')$ si, et seulement si, $Z = Z'$ et s'il existe une classe d'éléments h_i et une classe d'éléments g_i où $i \in I$, telles que : $h_i < (S, f, Z)$, $\beta(h_i) = \alpha(g_i)$, $g_i h_i < (S', f', Z')$, $\cup\alpha(h_i) = Z$, $h_i \in \mathfrak{H}$, $g_i \in \mathfrak{B}$.

Deux éléments h et h' de $\mathfrak{H}^-$ seront dits $\mathfrak{B}$-*compatibles* si les plus grands éléments induits par h et h' sur $\alpha(h) \cap \alpha(h')$ (c'est-à-dire les pseudoproduits $h(\alpha(h) \cap \alpha(h'))$ et $h'(\alpha(h) \cap \alpha(h'))$ (voir **5**)) sont équivalents modulo ρ.

On appellera *feuilletage de seconde espèce* $(\mathfrak{B}, \mathfrak{H})$ *élargie* un atlas complet de $\mathfrak{H}^\sim$ compatible avec $\mathfrak{B}$ et contenu dans $\mathfrak{H}^-$, c'est-à-dire une sous-classe A de $\mathfrak{H}^-$ vérifiant les conditions :

(1) A est saturé par induction et agrégation; $\cup\alpha(A)$ existe;

(2) Si $(S, f, Z) \in A$, $(S', f, Z) \in \mathfrak{H}$ et $S < S'$, alors $(S', f, Z) \in A$;

(3) La classe $\mathfrak{B}A$ des pseudoproduits ba, où $b \in \mathfrak{B}$ et $a \in A$, est identique à A;

(4) Deux éléments quelconques de A sont $\mathfrak{B}$-compatibles.

Les feuilletages de seconde espèce $(\mathfrak{B}, \mathfrak{H})$ élargie forment une espèce de structures locales $\mathfrak{A}(\mathfrak{B}, \mathfrak{H})_0$ étalée sur $\mathfrak{Z}_0$ par la projection $\zeta : A \to \cup\alpha(A)$.

$\mathfrak{Z}$ opère sur $\mathfrak{A}(\mathfrak{B}, \mathfrak{H})_0$, le composé de A avec (k, Z) étant l'atlas complet formé des pseudoproduits ak^{-1}, où $a \in A$, si, et seulement si, $Z = \zeta(A)$. Soit $\mathfrak{A}(\mathfrak{B}, \mathfrak{H})$ le groupoïde des isomorphismes correspondant.

Un feuilletage de seconde espèce $(\mathfrak{B}, \mathfrak{H})$ élargie peut aussi être défini par une classe complète de feuilletages élémentaires de la manière suivante :

Soit $\mathfrak{H}^-/\mathfrak{B}$ la classe quotient de $\mathfrak{H}^-$ par la relation d'équivalence r : $(S, f, Z) \sim (S', f', Z')$ si, et seulement si, il existe $(g, S) \in \mathfrak{B}$ tel que $(S', f', Z') = (g, S)(S, f, Z)$. La classe (S, f, Z) modulo r sera notée $\mathfrak{B}(S, f, Z)$.

$\mathfrak{H}^-/\mathfrak{B}$ est une classe sous-locale pour la relation : $h^\wedge < h^{\wedge\prime}$ si, et seulement si, il existe $h \in h^\wedge$ et $h' \in h^{\wedge\prime}$ avec $h < h'$, et l'application $\eta : \mathfrak{B}(S, f, Z) \to Z$ est un étalement de $\mathfrak{H}^-/\mathfrak{B}$ sur $\mathfrak{Z}_0$.

Soit r' la relation d'équivalence sur $\mathfrak{H}^-$: $(S, f, Z) \sim (S', f', Z')$ si, et seulement si, il existe $(S'', f'', Z'') \in \mathfrak{H}^-$ tel que $Z = Z' = Z''$, $f = f' = f''$, $S'' < S$ et $S'' < S'$.

Soit $\mathfrak{J}(\mathfrak{H}^-)$ la classe quotient de $\mathfrak{H}^-/\mathfrak{B}$ par la relation r'' : $h^\wedge \sim h^{\wedge\prime}$ si, et seulement si, il existe $h \in h^\wedge$ et $h' \in h^{\wedge\prime}$ tels que $h \sim h'$ modulo r'. Cette classe est une classe sous-locale pour l'ordre quotient de celui de $\mathfrak{H}^-/\mathfrak{B}$ par r'' et l'application η' :

$$\mathfrak{B}(S, f, Z) \text{ modulo } r'' \to Z$$

l'étale sur $\mathfrak{Z}_0$. Une classe $h^\sim \in \mathfrak{J}(\mathfrak{H}^-)$ sera appelée *feuilletage élémentaire de seconde espèce* sur $\eta'(h^\sim)$.

Soit C une classe complète de feuilletages élémentaires de seconde espèce, c'est-à-dire une sous-classe de $\mathfrak{J}(\mathfrak{H}^-)$ saturée par induction et agrégation et admettant pour base une sous-classe C' telle que, pour tout $c \in C'$ et tout $c' \in C'$, $c \cap c'$ soit défini et $\eta'(c) \cap \eta'(c') = \eta'(c' \cap c)$. Alors C s'identifie à un feuilletage de seconde espèce $(\mathfrak{B}, \mathfrak{H})$ élargie sur $Z = \cup \eta'(C)$.

Remarque. La catégorie $\mathfrak{H}^\sim$ peut être remplacée par une catégorie locale $\mathfrak{C}^\sim$ au-dessus de $\mathfrak{E}^\sim$ contenant $\mathfrak{B}$ et $\mathfrak{Z}$ comme sous-groupoïdes saturés par induction; il faut alors remplacer $\mathfrak{H}^-$ par la classe $\mathfrak{C}^-$ des éléments $h \in \mathfrak{C}^\sim$ tels que $\alpha(h) \in \mathfrak{Z}_0$ et $\beta(h) \in \mathfrak{B}_0$.

On pourrait aussi considérer la classe $\mathfrak{J}^\lambda(\mathfrak{H}^-)$ des jets locaux atomiques des éléments de $\mathfrak{H}^-$ au-dessus de $\mathfrak{E}^\sim$ (voir **5**). De la relation d'équivalence r, on déduit une relation d'équivalence dans $\mathfrak{J}^\lambda(\mathfrak{H}^-)$ dont les classes d'équivalence sont appelées *germes de feuilletages* (voir **4** et **16**); la classe des germes de feuilletages est un faisceau sur l'espace (méta-) topologique des germes atomiques de structures de l'espèce $\mathfrak{Z}_0$, dont les sections au-dessus de $Z \in \mathfrak{Z}_0$ sont les feuilletages de seconde espèce $(\mathfrak{B}, \mathfrak{H}^-)$ élargie.

Exemples

(1) Soit $\mathfrak{C}^r$ la catégorie de tous les homomorphismes r fois différentiables, $\mathfrak{B}$ étant le groupoïde $\Lambda_p{}^r$ et $\mathfrak{Z}$ le groupoïde des automorphismes locaux d'ordre r d'une variété différentiable V_n de dimension n. Un atlas complet A compatible avec $\Lambda_p{}^r$ formé d'homomorphismes r fois différentiables d'ouverts de V_n dans R^p sera

appelé *feuilletage r fois différentiable de seconde espèce*. En particulier, si A est formé d'homomorphismes r fois différentiables de rang constant, A est appelé feuilletage r fois différentiable *régulier* A; est sous-jacent à une structure de variété feuilletée r fois différentiable bien déterminée (voir ci-dessous). Les singularités d'un feuilletage A r fois différentiable de seconde espèce correspondent aux germes de A provenant d'un homomorphisme h de rang non constant au voisinage d'un point.

(2) Soit $\mathfrak{Z} = \mathfrak{L}(\mathfrak{B} \# \mathfrak{F})$ (voir paragraphe 4); soit $\mathfrak{H}$ la classe des triplets $(s, pr_1 f^{-1}, \sigma)$, où $\sigma = (f, s \times s')\mathfrak{B} \# \mathfrak{F}$ et pr_1 est la projection canonique de $p(s) \times p'(s')$ sur $p(s)$. Alors $\mathfrak{H}$ vérifie les conditions 1, 2, 3 du début du paragraphe et on peut lui associer la classe $\mathfrak{H}^-(\mathfrak{B} \# \mathfrak{F})$ des homomorphismes de $\mathfrak{L}(\mathfrak{B} \# \mathfrak{F})$ vers $\mathfrak{B}$. A une structure feuilletée Σ de l'espèce $\mathfrak{B} \# \mathfrak{F}$ élargie correspond l'atlas complet A de $\mathfrak{H}^-(\mathfrak{B} \# \mathfrak{F})$ engendré par les cartes $(s, pr_1 f^{-1}, \sigma)$, où $\sigma = (f, s \times s')\mathfrak{B} \# \mathfrak{F} \in \mathfrak{Z}$; cet atlas complet définit un feuilletage de seconde espèce $(\mathfrak{B}, \mathfrak{H}(\mathfrak{B} \# \mathfrak{F}))$ élargie sous-jacent à Σ.

Soit $A \in \mathfrak{A}(\mathfrak{B}, \mathfrak{H})_0$ un feuilletage de seconde espèce $(\mathfrak{B}, \mathfrak{H})$ élargie sur $Z \in \mathfrak{Z}_0$. Soit $\tau(Z)$ la topologie sur $p_1(Z)$ sous-jacente à Z. Les traces sur les ouverts de $\tau(Z)$ des ensembles $f^{-1}(x)$, où $(s, f, z) \in A$ et $x \in p(s)$, forment une base d'une topologie T' sur $p_1(Z)$. Le couple des topologies $(\tau(Z), T')$ définit sur $p_1(Z)$ un feuilletage topologique sous-jacent au feuilletage de seconde espèce A. Si T' est localement connexe, une feuille de $(\tau(Z), T')$ sera appelée feuille de A.

Si A est le feuilletage de seconde espèce sous-jacent à la structure feuilletée $\Sigma \in \mathfrak{L}(\mathfrak{B} \# \mathfrak{F})_0$, les feuilletages topologiques sous-jacents à Σ et à A sont identiques et ce sont des feuilletages localement simples.

Soit (T, T') un feuilletage topologique localement simple sur E; pour tout ouvert simple U de E, soit $U^\vee$ son espace transverse et u l'application canonique de U sur $U^\vee$. Soit $\mathfrak{T}$ le groupoïde des homéomorphismes d'un espace topologique sur un autre et soit $\mathfrak{Z}$ la classe des ouverts de E relativement à T. Soit $\mathfrak{H}^-$ la classe de toutes les applications continues d'un élément de $\mathfrak{Z}$ dans un espace topologique. La sous-classe de $\mathfrak{H}^-$ formée des triplets $(U^\vee, u, U)$ engendre un atlas complet A compatible avec $\mathfrak{T}$ qui définit sur E un feuilletage de seconde espèce $(\mathfrak{T}, \mathfrak{H}^-)$ élargie sous-jacent au feuilletage topologique (T, T'). Le feuilletage topologique sous-jacent à A est aussi le feuilletage (T, T').

II. Structure transversale d'un feuilletage

1. *Pseudogroupe d'holonomie*

Soient (T, T') un feuilletage topologique localement simple sur E et U un ouvert simple.

Soit ν un relèvement d'un sous-espace B de l'espace transverse $U^{\check{}}$ dans U, c'est-à-dire une application continue de B dans U (muni de la topologie induite par T) telle qui $\nu(x^{\check{}})$ appartienne à la plaque de U ayant $x^{\check{}}$ pour image canonique dans U. Le sous-espace $\nu(B)$ de U sera appelé *espace quasi-transversal élémentaire.* Si $B = U^{\check{}}$, $\nu(B)$ sera appelé *espace transversal élémentaire.*

Les germes des espaces quasi-transversaux élémentaires forment un ouvert $E_\kappa^{\hat{}}$ de l'espace $E^{\hat{}}$ de tous les germes de sous-espaces de l'espace E muni de la topologie T. Muni de la topologie induite par celle de $E^{\hat{}}$, l'espace $E_\kappa^{\hat{}}$ sera appelé *espace quasi-transversal total.* Le sous-espace de $E_\kappa^{\hat{}}$ formé des germes d'espaces transversaux élémentaires sera appelé *espace transversal total* de (T, T') et noté $E_\theta^{\hat{}}$. Un ouvert supposé relativement séparé de $E_\kappa^{\hat{}}$ (respectivement de $E_\theta^{\hat{}}$) sera appelé *espace quasi-transversal* (respectivement *transversal*). En vertu du théorème de Zorn, tout espace (quasi-)transversal est contenu dans un espace (quasi-)transversal maximal.

La classe Θ des couples (h, N), où N est un espace transversal élémentaire et h un homéomorphisme de N sur un espace transversal élémentaire N', est un sous-groupoïde local de $\mathfrak{T}$ et un groupoïde local au-dessus de $\mathfrak{E}$ pour la projection π: $(h, N) \rightarrow (h, \pi(N))$, où $\pi(N)$ est l'ensemble sous-jacent au sous-espace N. Un élément de Θ sera appelé *isomorphisme transversal élémentaire* et on appellera *isomorphisme localement transversal* une classe complète C de Θ compatible avec π. Si C est formé d'unités (c'est-à-dire d'espaces transversaux élémentaires), C définit un espace transversal sans point double. Θ contient les sous-groupoïdes suivantes :

(1) Le groupoïde des couples (h, N) tels que, pour tout $x \in \pi(N)$, x et $h(x)$ appartiennent à une même feuille de (T, T'). Un élément de ce sous-groupoïde Θ_B sera appelé *isomorphisme transversal strict élémentaire.*

(2) Le groupoïde H engendré par la classe H_s des (h, N) tels que $N = \nu(U^{\check{}})$, $h(\pi(N)) \subset U$ et $uh\nu =$ Identité de $U^{\check{}}$, où u est l'application canonique de U sur $U^{\check{}}$. Le pseudogroupe H sera appelé *pseudogroupe d'holonomie.*

(3) Pour tout $x \in E$, le groupoïde H_x des couples $(h, N) \in H_s$ tels que $\pi(N) \cap h(\pi(N))$ contienne le point x.

Soit $\Theta^{\wedge}$ le groupoïde des jets locaux $j_x^{\lambda}(h, N)$, où $(h, N) \in \Theta$ et $x \in \pi(N)$; $\Theta^{\wedge}$ admet $E_\theta^{\wedge}$ pour classe d'objets. Soit $h^{\wedge}$ l'ensemble des jets locaux $j_x^{\lambda}(h, N)$, où x décrit $\pi(N)$; les ensembles $h^{\wedge}$ forment une base d'une topologie sur $\Theta^{\wedge}$, induisant sur $E_\theta^{\wedge}$ la topologie considérée ci-dessus. Pour cette topologie, $\Theta^{\wedge}$ est un groupoïde topologique (voir **7**).

Soient $\Theta_s^{\wedge}$ (respectivement $H^{\wedge}$, $H_x^{\wedge}$) les sous-groupoïdes de $\Theta^{\wedge}$ formés des jets $j_x^{\lambda}(h, N)$, où $(h, N) \in \Theta_s$ (respectivement $\in H$, $\in \mathrm{H}_x$); ces sous-groupoïdes sont ouverts dans $\Theta^{\wedge}$.

Un ouvert Ω de $\Theta^{\wedge}$, tel que la correspondance définie par les couples $(\alpha(X), \beta(X))$, où $X \in \Omega$, soit une application biunivoque, sera appelé *isomorphisme transversal*; Ω s'identifie à un homéomorphisme de l'espace transversal $\alpha(\Omega)$ sur l'espace transversal $\beta(\Omega)$. Les éléments Ω forment un pseudogroupe Θ'; H s'identifie à un sous-pseudogroupe faible de Θ'; le sous-pseudogroupe $\tilde{H}$ de Θ' ayant pour base H a pour éléments les isomorphismes transversaux Ω contenus dans $H^{\wedge}$.

Le groupoïde $\tilde{\tilde{H}}$ quotient de $H^{\wedge}$ par le groupoïde réunion des groupoïdes $H_x^{\wedge}$ est appelé *groupoïde d'holonomie* de (T, T') et un élément de $\tilde{\tilde{H}}$, *jet d'holonomie.* La classe d'équivalence X de $\eta = j_x^{\lambda}(h, N) \in H^{\wedge}$ est formée des jets locaux $s'\eta s$, où $s \in H_x^{\wedge}$, $s' \in H_{x'}^{\wedge}$ et $x' = h(x)$; les unités à droite et à gauche de X sont respectivement $H_x^{\wedge}$ et $H_{x'}^{\wedge}$; de plus, on posera : $\alpha^{\wedge}(X) = x$ et $\beta^{\wedge}(X) = x'$.

On supposera généralement que par tout point de E passe au moins un espace transversal. Soit F une feuille de (T, T'); soit Φ_x la classe des jets d'holonomie X tels que $\alpha^{\wedge}(X) = x$, où $x \in F$; alors on a : $\beta^{\wedge}(\Phi_x) = F$.

2. *Groupoïde transverse d'holonomie*

Soit (T, T') un feuilletage topologique localement simple sur

E; soit $\mathfrak{I}$ le groupoïde des couples (h, U), où U est un ouvert simple de (T, T') et h un isomorphisme de $(T, T')_U$ sur $(T, T')_{U'}$. Alors $\mathfrak{I}$ s'identifie à un sous-pseudogroupe de $\mathfrak{T}$ et $\mathfrak{I}$ contient les sous-pseudogroupes suivants :

(1) Le groupoïde $\mathfrak{I}_s$ des couples (h, U) tels que, pour tout $x \in U$, x et $h(x)$ appartiennent à une même feuille de (T, T').

(2) Le groupoïde $\mathfrak{I}_c$ engendré par les couples (h, U), où U et $h(U)$ sont des ouverts distingués d'un même ouvert simple U_1 et où x et $h(x)$ appartiennent à une même plaque de U_1 pour tout $x \in U$.

A $(h, U) \in \mathfrak{I}$ correspond une application $h^{\vee}$ de $U^{\vee}$ sur $U^{\vee\prime}$, où $U^{\vee}$ et $U^{\vee\prime}$ sont les espaces transverses de $(T, T')_U$ et de $(T, T')_{U'}$ respectivement et $U' = h(U)$.

Soit $\mathfrak{I}/\rho$ le groupoïde ordonné quotient **(voir 3)** de $\mathfrak{I}$ par la relation d'équivalence : $(h, U) \sim (h_1, U_1)$ si, et seulement si, $U = U_1$, $h(U) = h_1(U_1)$ et $h_1^{\vee} = h^{\vee}$. La classe d'équivalence de $(h, U) \in \mathfrak{I}$ sera identifiée au triplet $(U', h^{\vee}, U)$, où $U' = h(U)$, et $\mathfrak{I}/\rho$ s'identifie à un sous-groupoïde du groupoïde $\mathfrak{I}'$ suivant : $\mathfrak{I}'$ est le groupoïde des triplets (U', f, U), où f est un homéomorphisme de l'espace transverse $U^{\vee}$ sur l'espace transverse $U^{\vee\prime}$. On munit $\mathfrak{I}'$ de la relation d'ordre : $(U', f, U) < (U_1', f_1, U_1)$ si, et seulement si, U est distingué dans U_1, U' est distingué dans U_1' et $u'f = f_1 u$, où u et u' sont les injections canoniques de $U^{\vee}$ et $U^{\vee\prime}$ dans $U_1^{\vee}$ et $U_1^{\vee\prime}$ respectivement. Un élément de $\mathfrak{I}'$ sera appelé *isomorphisme transverse.*

$\mathfrak{I}'$ admet pour sous-groupoïde le groupoïde H' des triplets (U', f, U) tels que f soit l'isomorphisme canonique associé à une chaîne pure (voir § I.1); H' est engendré par les triplets (U', f, U) tels que U soit distingué dans U' et que f soit l'injection de $U^{\vee}$ dans $U^{\vee\prime}$, supposée être une surjection. H', muni de la structure d'ordre induite par $\mathfrak{I}'$, est appelé *groupoïde des isomorphismes d'holonomie.*

Remarque. On pourrait aussi munir $\mathfrak{I}'$ de la relation d'ordre : $(U', f, U) < (U_1', f_1, U_1)$ si, et seulement si, $U \subset U_1$, $U' \subset U_1'$, $u'f = f_1 u$; pour cette relation, $\mathfrak{I}'$ est une catégorie locale (dans laquelle un élément induit par une unité peut ne pas être une unité) dont $\mathfrak{I}/\rho$ est une sous-catégorie locale et H' une classe prélocale **(3)**.

Soient U et U' deux ouverts simples de E. Soit $H'(U', U)$ la sous-classe de H' formée des triplets (U_1', f_1, U_1), où U_1' et U_1 sont des ouverts saturés de U' et U respectivement; alors $H'(U', U)$ est une classe locale appelée *classe transverse d'holonomie relative à* (U', U). Soient $U^\smile$ et $U^{\smile\prime}$ les espaces transverses de U et de U'; l'application : $(U_1', f_1, U_1) \to (U_1'^\smile, f_1, U_1^\smile)$ identifie $H(U', U)$ à une classe $\Phi(U', U)$ d'homéomorphismes d'un sous-espace de $U^\smile$ sur un sous-espace de $U^{\smile\prime}$, appelés homéomorphismes d'holonomie de $U^\smile$ vers $U^{\smile\prime}$. En particulier, si $U = U'$, on peut identifier $H'(U, U)$ et $\Phi(U, U)$; on obtient ainsi un pseudogroupe que l'on désignera par $\Phi(U)$ et appellera *pseudogroupe transverse d'holonomie relatif à* U. Le groupoïde des jets locaux des applications de $\Phi(U)$ sera appelé *groupoïde transverse d'holonomie relatif à* U.

Soient $x \in E$ et $x' \in E$; designons par ${}_{x'}\mathfrak{F}_x'$ la classe des éléments $(U', f, U) \in \mathfrak{F}'$ où $x^{\smile\prime} = f(x^\smile)$, $x \in U$ et $x' \in U'$; soit ${}_{x'}\rho_x$ la relation d'équivalence dans ${}_{x'}\mathfrak{F}_x'$: $(U', f, U) \sim (U_1', f_1, U_1)$ si, et seulement si, il existe $(U_2', f_2, U_2) \in {}_{x'}\mathfrak{F}_x'$ tel que $(U_2', f_2, U_2) < (U', f, U)$ et $(U_2', f_2, U_2) < (U_1', f_1, U_1)$. La classe (U', f, U) modulo ${}_{x'}\rho_x$ sera désignée par ${}_{x'}j_x^\lambda(U', f, U)$ et appelée *jet local transverse.* Par passage au quotient relativement à la relation d'équivalence engendrée par les relations ${}_{x'}\rho_x$ on obtient le groupoïde $J^\lambda(\mathfrak{F}')$ des jets locaux transverses. Le sous-groupoïde $J^\lambda(H')$ de $J^\lambda(\mathfrak{F}')$ formé des jets ${}_{x'}j_x^\lambda(U', f, U)$, où $(U', f, U) \in H'$, sera appelé *groupoïde d'holonomie* et noté $\tilde{H}'$.

PROPOSITION. *Si par tout point x de E il passe au moins un espace transversal, $H^{\sim\prime}$ est identique au groupoïde d'holonomie $H^\sim$* (§ 1).

Si $X = (U', f, U) \in H'$, soit $W(X)$ l'ensemble des éléments ${}_{x_1'}j_{x_1}^\lambda(U', f, U) \in H^{\sim\prime}$, où $x_1 \in U$ et $x_1' \in U'$. Les ensembles $W(X)$ forment une base d'une topologie sur $H^{\sim\prime}$.

PROPOSITION. *Muni de la topologie précédente, $H^{\sim\prime}$ est un groupoïde topologique* (voir **7**).

Pour tout $x \in E$, le sous-groupe $H_x^{\sim\prime}$ de $H^{\sim\prime}$ formé des jets locaux ${}_xj_x^\lambda(U', f, U)$ sera appelé *groupe d'holonomie* en x de la feuille F passant par x. L'unité de $H_x^{\sim\prime}$, appelée *germe d'espace transverse en* x, s'identifie à l'ensemble des couples (U, ρ_U), où U est un ouvert simple contenant x et ρ_U la relation d'équivalence sous-jacente à

$(T, T')_U$ (c'est-à-dire le germe en x de la partition locale sous-jacente).

3. *Déroulements d'un espace feuilleté*

Soit (T, T') un feuilletage localement simple sur E.

Soient ω un ouvert simple de E et $\omega^{\check{}}$ son espace transverse. Soit $X = (U', f, U) \in H'$, où U' est un ouvert saturé dans ω pour la relation d'équivalence ρ_ω sous-jacente à $(T, T')_\omega$. Soit $x^{\check{}\prime}$ une plaque de U'; soit $\xi = {}_{x^{\check{}\prime}}j_x^{\lambda}(U', f, U)$ la classe des triplets $(U_l', f_l, U_l) \in H'$ vérifiant les conditions suivantes : U_1' est saturé pour ρ_ω; il existe $(U_2', f_2, U_2) \in H'$ tel que U_2' est saturé pour ρ_ω, $(U_2', f_2, U_2) < (U', f, U)$, $(U_2', f_2, U_2) < (U_1', f_1, U_1)$, $x \in U_2$, $x^{\check{}\prime}$ est la plaque de U_2' contenant $f_2u_2(x)$, où u_2 est l'application canonique de U_2 sur $U_2^{\check{}}$. On peut identifier ξ avec $j_x^{\lambda}(ifu)$, jet local de E dans $\omega^{\check{}}$, où u et i sont les applications canoniques de U sur $U^{\check{}}$ et de $U^{\check{}\prime}$ dans $\omega^{\check{}}$. Soit $E^*(\omega)$ la classe des jets ξ ainsi définis. Soit $W(X)$ l'ensemble des jets locaux ${}_{x_1^{\check{}\prime}}j_{x_1}^{\lambda}(U', f, U)$, où x_1' décrit U' et x_1 décrit U. Les ensembles $W(X)$ forment une base d'une topologie T^* sur $E^*(\omega)$. Désignons par Ω la réunion des feuilles F de (T, T') rencontrant ω. Soit α^* l'application : $\xi \to x$ de $E^*(\omega)$ dans Ω et β^* l'application $\xi \to x^{\check{}\prime}$ de $E^*(\omega)$ dans $\omega^{\check{}}$.

PROPOSITION. *Avec les notations précédentes, $E^*(\omega)$ est étalé au-dessus de Ω par la projection α^* et le feuilletage $(T^*, T^{*\prime})$ déterminé sur $E^*(\omega)$ par α^* est simple. La feuille $F^* = \beta^{*-1}(x^{\check{}\prime})$, où $x' \in \omega$, est un revêtement normal de $F \ni x'$ pour la projection α^*; le groupe des automorphismes de ce revêtement s'identifie au groupe d'holonomie $\tilde{H}_{x'}'$, en associant à $s \in \tilde{H}_{x'}'$ l'automorphisme $\xi \to s\xi$, où $\xi \in F^*$.*

On appelle $E^*(\omega)$ le *déroulement faible* de (T, T') au-dessus de ω. La topologie de $E^*(\omega)$ peut ne pas être séparée même si T est séparée.

Soit $\mathfrak{C}$ la classe des chaînes pures munie de la loi de composition définie par :

$$(W_1, \ldots, W_n; W_{1,2}, \ldots, W_{n-1,n})(W_1', \ldots, W_{n'}'; W_{1,2}', \ldots, W_{n'-1,n'})$$
$$= (W_1, \ldots, W_n, W_2', \ldots, W_{n'}'; W_{1,2}, \ldots, W_{n-1,n}, W_{1,2}', \ldots, W_{n'-1,n'}')$$

si, et seulement si, $W_1' = W_n$.

$\mathfrak{C}$ est une catégorie libre dont les unités sont les chaînes pures réduites à un seul élément (W).

Soit r la plus petite relation d'équivalence sur $\mathfrak{C}$ compatible avec la loi de composition et avec les relations élémentaires :

$$(W) \sim (W, W; W); \quad (W_1, W_2; W_{1,2})(W_2, W_1, W_{1,2}) \sim (W_1).$$

La classe d'équivalence Γ modulo r sera notée $[\Gamma]$. La classe $\mathfrak{C}/r$ est un groupoïde pour la loi de composition déduite de celle de $\mathfrak{C}$ par passage au quotient. $\mathfrak{C}/r$ s'identifie au groupoïde libre associé au graphe dont les sommets sont les chaînes pures (W) et les arêtes les éléments $(W_1, W_2; W_{1,2})$.

La catégorie $\mathfrak{C}$ est munie de la relation d'ordre : $(W_1, \ldots, W_n; W_{1,2}, \ldots, W_{n-1,n}) < (W_1', \ldots, W_n'; W_{1,2}', \ldots, W_{n'-1,n'}')$ si, et seulement si, $n = n'$, W_i est distingué dans W_i' et la plaque de $W_{i,i+1}$ déterminée par x est contenue dans la plaque de $W_{i',i+1}'$ déterminée par x.

Soit $\mathfrak{C}^{\cdot}$ la classe des chaînes pures pointées; $\mathfrak{C}^{\cdot}$ est une catégorie pour la loi de composition : $(x, \Gamma, x')(x_1', \Gamma', x'') = (x, \Gamma\Gamma', x'')$ si, et seulement si, $x' = x_1'$ et le composé $\Gamma\Gamma'$ est défini dans $\mathfrak{C}$. Soit $\mathfrak{C}^{\cdot}/r$ le groupoïde quotient de $\mathfrak{C}^{\cdot}$ par la relation d'équivalence $r^{\cdot}$: $(x, \Gamma, x') \sim (x_1, \Gamma_1, x_1')$ si, et seulement si, $x = x_1$, $x' = x_1'$ et $\Gamma \sim \Gamma_1$ modulo r. La classe (x, Γ, x') modulo $r^{\cdot}$ sera notée $[x, \Gamma, x']$.

$\mathfrak{C}^{\cdot}$ est munie de la relation d'ordre : $(x, \Gamma, x') < (x_1, \Gamma_1, x_1')$ si, et seulement si, $x = x_1$, $x' = x_1'$ et $\Gamma < \Gamma_1$; par passage au quotient, $\mathfrak{C}^{\cdot}/r$ est aussi muni d'une relation de préordre.

Soit r' la plus petite relation d'équivalence sur $\mathfrak{C}^{\cdot}/r$ qui soit compatible avec la loi de composition et avec la relation : $C \sim C'$ si, et seulement si, $C < C'$ ou si $C' < C$. Nous désignerons par $\mathfrak{C}^{\sim}$ le groupoïde quotient de $\mathfrak{C}^{\cdot}/r$ par r'.

Soit $\mathfrak{R}$ un recouvrement de E par des ouverts simples U. Soit $\mathfrak{C}_{\mathfrak{R}}^{\cdot}$ la sous-catégorie de $\mathfrak{C}^{\cdot}$ formée des chaînes pures pointées $(x, W_1, \ldots, W_n; W_{1,2}, \ldots, W_{n-1,n}, x')$ telles que W_i soit un ouvert saturé dans un ouvert $U \in \mathfrak{R}$. Soit $\mathfrak{C}_{\mathfrak{R}}^{\cdot}/r$ le sous-groupoïde de $\mathfrak{C}^{\cdot}/r$ quotient de $\mathfrak{C}_{\mathfrak{R}}^{\cdot}$ par la relation d'équivalence $\dot{r}_{\mathfrak{R}}$ induite par $\dot{r}$ sur $\mathfrak{C}_{\mathfrak{R}}^{\cdot}$; cette relation d'équivalence est aussi la plus petite relation d'équivalence compatible avec la loi de composition et avec les relations élémentaires entre éléments de $\mathfrak{C}_{\mathfrak{R}}^{\cdot}$. Soit $r_{\mathfrak{R}}'$ la plus petite relation d'équivalence sur $\mathfrak{C}_{\mathfrak{R}}^{\cdot}/r$ compatible avec la loi de composition et avec la relation : $C \sim C'$ si, et seulement si, il existe $C'' \in \mathfrak{C}^{\cdot}/r$ tel que $C'' < C$ et $C'' < C'$. Pour tout $c \in \mathfrak{C}_{\mathfrak{R}}^{\cdot}$, la classe ($c$ modulo $\dot{r}\mathfrak{R}$) modulo $r\mathfrak{R}'$ sera notée $[c]$ modulo $r\mathfrak{R}'$. Soit $\mathfrak{C}\mathfrak{R}^{\sim}$ le groupoïde quotient de $\mathfrak{C}\mathfrak{R}'/r$ par $r\mathfrak{R}'$. L'application : $[c]$ modulo $r\mathfrak{R}' \to [c]$modulo r' est une application de $\mathfrak{C}\mathfrak{R}^{\sim}$ sur $\mathfrak{C}^{\sim}$, qui n'est généralement pas biunivoque.

Pour tout $\Gamma = (W_1, \ldots, W_n; W_{1,2}, \ldots, W_{n-1,n}) \in \mathfrak{C}\mathfrak{R}$, soit $U(\Gamma)$ l'ensemble des éléments $\tilde{C} = [x, \Gamma, x']$ modulo $r\mathfrak{R}' \in \mathfrak{C}\mathfrak{R}^{\sim}$, où $x \in W_1$ et $x' \in W_n$. L'ensemble $U(\Gamma)$ sera muni de la topologie $T(\Gamma)$ ayant pour base les ensembles $U(\Gamma, W_1', W_n')$ des éléments $[x, \Gamma, x']$ modulo $r\mathfrak{R}'$ où $x \in W_1'$, $x' \in W_n'$, W_1' et W_n' étant des ouverts contenus dans W_1 et W_n. On munit aussi $U(\Gamma)$ de la topologie $T'(\Gamma)$ ayant pour base les ensembles $U(\Gamma, V_1', V_n')$ des classes $[x, \Gamma, x']$ modulo $r\mathfrak{R}'$, où $x \in V_1'$, $x' \in V_n'$, V_1' et V_n' étant des plaques de W_1' et W_n'. Le couple $(T(\Gamma), T'(\Gamma))$ définit un feuilletage topologique sur $U(\Gamma)$.

PROPOSITION. *Avec les notations précédentes, $\mathfrak{C}\mathfrak{R}^{\sim}$ est un groupoïde topologique pour la topologie $T\mathfrak{R}$ (respectivement $T\mathfrak{R}'$) ayant pour base les ensembles $U(\Gamma, W_1', W_n')$ (respectivement $U(\Gamma, V_1', V_n')$) et le couple $(T\mathfrak{R}, T\mathfrak{R}')$ est un feuilletage topologique.*

Le sous-espace de $\mathfrak{C}\mathfrak{R}^{\sim}$ formé des unités s'identifie à E muni de la topologie T (respectivement T').

Nous désignerons par $\alpha^{\sim}$ et $\beta^{\sim}$ les projections :

$$\alpha^{\sim} : [x, \Gamma, x'] \text{ modulo } r\mathfrak{R}' \to x'$$
$$\beta^{\sim} : [x, \Gamma, x'] \text{ modulo } r\mathfrak{R}' \to x$$

de $\mathfrak{C}\mathfrak{R}^{\sim}$ sur E.

Soit $\mathfrak{N}$ un espace transversal; désignons par $\tilde{E}\mathfrak{R}(\mathfrak{N})$ le sous-espace de $\mathfrak{C}\mathfrak{R}^{\sim}$ formé des éléments $\tilde{C}$ tels que $\beta^{\sim}(\tilde{C}) \in \mathfrak{N}$. Alors la projection $\alpha^{\sim}$ étale $\tilde{E}\mathfrak{R}(\mathfrak{N})$ sur le sous-espace de E réunion des feuilles de (T, T') qui rencontrent $\mathfrak{N}$. On appellera $\tilde{E}\mathfrak{R}(\mathfrak{N})$ le *déroulement relatif à* $\mathfrak{R}$ *au-dessus de* $\mathfrak{N}$. Le feuilletage $(\tilde{T}\mathfrak{R}, \tilde{T}\mathfrak{R}')$ induit sur $\tilde{E}\mathfrak{R}(\mathfrak{N})$ par $(T\mathfrak{R}, T\mathfrak{R}')$ s'identifie au feuilletage défini par l'étalement $\alpha^{\sim}$.

PROPOSITION. *Avec les notations précédentes, le feuilletage $(\tilde{T}\mathfrak{R}, \tilde{T}\mathfrak{R}')$ de $\tilde{E}\mathfrak{R}(\mathfrak{N})$ est simple et son espace transverse est identifié à $\mathfrak{N}$ par l'application $\beta^{\sim}$.*

Soit F une feuille de (T, T'), $\mathfrak{R}(F)$ le recouvrement de F par les plaques V des ouverts $U \in \mathfrak{R}$. Soit $\mathfrak{C}_{\mathfrak{R}(F)}^{\cdot}$ l'ensemble des chaînes de plaques pointées $(x, \gamma, x') = (x, V_1, \ldots, V_n; V_{1,2}, \ldots, V_{n-1,n}, x')$ où $V \in \mathfrak{R}(F)$ pour tout $i \leqslant n$. On définit comme précédemment (en considérant sur F le feuilletage trivial (T', T')) une loi de composition dans $\mathfrak{C}_{\mathfrak{R}(F)}^{\cdot}$ et des relations $\dot{r}_{\mathfrak{R}(F)}$ et $r_{\mathfrak{R}(F)}'$. Soit $\mathfrak{C}_{\mathfrak{R}(F)}^{\sim} = (\mathfrak{C}_{\mathfrak{R}(F)}^{\cdot}/r_{\mathfrak{R}(F)})/r_{\mathfrak{R}(F)}'$. Pour tout élément (x, γ, x') de $\mathfrak{C}_{\mathfrak{R}(F)}^{\cdot}$ il existe des chaînes pures pointées appartenant à $\mathfrak{C}_{\mathfrak{R}}^{\cdot}$ dont (x, γ, x') soit la chaîne de plaques associée reliant x à x'; soit $C(x, \gamma, x')$ une de ces chaînes. Soit φ l'application :

$$[x, \gamma, x'] \text{ modulo } r_{\mathfrak{R}(F)}' \to [C(x, \gamma, x')] \text{ modulo } r_{\mathfrak{R}}'.$$

PROPOSITION. *Avec les notations précédentes, φ est un isomorphisme de $\mathfrak{C}_{\mathfrak{R}(F)}^{\sim}$ sur le sous-groupoïde de $\mathfrak{C}_{\mathfrak{R}}^{\sim}$ formé des éléments $\tilde{C}$ tels que $\beta^{\sim}(\tilde{C}) = x$, où x décrit F.*

Supposons x fixé dans F. Soit $\tilde{E}_{\mathfrak{R}(F)}(x)$ le sous-espace topologique de $\mathfrak{C}_{\mathfrak{R}(F)}^{\sim}$ formé des éléments $[x, \gamma, x']$ modulo $r_{\mathfrak{R}(F)}'$ et $\mathfrak{C}_{\mathfrak{R}}^{\sim}(x)$ le sous-groupe de $\mathfrak{C}_{\mathfrak{R}(F)}^{\sim}$ formé des éléments $[x, \gamma, x]$ modulo $r_{\mathfrak{R}(F)}'$; $\mathfrak{C}_{\mathfrak{R}}^{\sim}(x)$ s'identifie aussi à un sous-groupe de $\mathfrak{C}_{\mathfrak{R}}^{\sim}$.

PROPOSITION. *Avec les notations précédentes, $\tilde{E}_{\mathfrak{R}(F)}(x)$ est un revêtement normal de F pour l'application : $[x, \gamma, x']$ modulo $r_{\mathfrak{R}(F)}' \to x'$; $\mathfrak{C}_{\mathfrak{R}}^{\sim}(x)$ est antiisomorphe au groupe des automorphismes de ce revêtement, Si $\mathfrak{R}(F)$ est formé d'ouverts simplement connexes, $\tilde{E}_{\mathfrak{R}(F)}(x)$ s'identifie au revêtement universel de F et $\mathfrak{C}_{\mathfrak{R}}^{\sim}(x)$ au groupe fondamental de F.*

L'antiisomorphisme est l'application qui associe à $s \in \mathfrak{C}_{\mathfrak{R}}^{\sim}(x)$ l'automorphisme du revêtement $y \to sy$, où $y \in \tilde{E}_{\mathfrak{R}(F)}(x)$.

Soient $C = (x, \Gamma, x') \in \mathfrak{C}_{\mathfrak{R}}^{\cdot}$ et h l'isomorphisme canonique de $W_1^{\smile}$ sur $W_n^{\smile}$ associé à C (voir § I, 1); désignons par $j^{\lambda}(C)$ le jet local ${}_{x^{\smile}}j_x^{\lambda}(W_n, h, W_1) \in \tilde{H}'$. Soit $\tilde{C} = [C]$ modulo $r_{\mathfrak{R}}' = [C_1]$ modulo $r_{\mathfrak{R}}'$ où $C_1 \in \mathfrak{C}_{\mathfrak{R}}^{\cdot}$; alors on a $j^{\lambda}(C) = j^{\lambda}(C_1)$. Soit ψ l'application : $\tilde{C} \to j^{\lambda}(C)$.

PROPOSITION. *Avec les notations précédentes, ψ est un foncteur contravariant de $\mathfrak{C}_{\mathfrak{R}}^{\sim}$ sur $\tilde{H}'$. En particulier, on a une antireprésentation de $\mathfrak{C}_{\mathfrak{R}}^{\sim}(x)$ sur le groupe d'holonomie $\tilde{H}x'$, où $x \in F$.*

Soit $W \in \mathfrak{R}$. Soit $\mathfrak{C}_{\mathfrak{R}}^{\sim}(W)$ l'ensemble des éléments $[x, \Gamma, x']$ modulo $r_{\mathfrak{R}}'$ tels que le premier élément W_1 de Γ soit un ouvert saturé de W; cet ensemble est muni de la relation d'équivalence $r\mathfrak{R}''$: $[x, \Gamma, x']$modulo $r\mathfrak{R}' \sim [x_1, \Gamma_1, x_1']$ modulo $r\mathfrak{R}'$ si, et seulement si, $x' = x_1'$, $\Gamma = \Gamma_1$ et x_1 appartient à la plaque $x^{\check{}}$ de W contenant x. La classe d'équivalence ($[x, \Gamma, x']$modulo $r\mathfrak{R}'$)modulo $r\mathfrak{R}''$ sera notée $[x^{\check{}}, \Gamma, x']_{r'(\mathfrak{R})}$. Soit $\tilde{E}_{\mathfrak{R}}(W)$ l'ensemble des classes d'équivalence modulo $r\mathfrak{R}''$.

Soit Γ une chaîne pure $(W_1, \ldots, W_n; W_{1,2}, \ldots, W_{n-1,n})$, où W_1 est saturé dans W, et soit $U(\Gamma, W_1', W_n')$ l'ensemble des éléments $[x^{\check{}}, \Gamma, x']_{r'(\mathfrak{R})}$, où $x^{\check{}}$ est une plaque de W_1', $x' \in W_n$, W_1' est un ouvert saturé dans W_1 et W_n' un ouvert contenu dans W_n; soit $U(\Gamma, x^{\check{}}, V_n')$ l'ensemble des éléments $[x^{\check{}}, \Gamma, x']_{r'(\mathfrak{R})}$, où $x' \in V_n'$, V_n' étant une plaque de W_n'. Soit $T_{\mathfrak{R}}(W)$ (respectivement $T_{\mathfrak{R}}'(W)$) la topologie sur $\tilde{E}_{\mathfrak{R}}(W)$ ayant pour base les ensemble $U(\Gamma, W_1', W_n')$ (respectivement $U(\Gamma, x^{\check{}}, V_n')$). Le couple $(T_{\mathfrak{R}}(W), T_{\mathfrak{R}}'(W))$est un feuilletage topologique simple dont l'espace transverse s'identifie à l'espace transverse $W^{\check{}}$ de W par l'application :

$$[x^{\check{}}, \Gamma, x']_{r'(\mathfrak{R})} \to x^{\check{}}.$$

De plus l'application $\alpha^{\sim}$: $[x^{\check{}}, \Gamma, x']_{r'(\mathfrak{R})} \to x'$ est un étalement de $\tilde{E}_{\mathfrak{R}}(W)$ sur la réunion des feuilles de (T, T') rencontrant W. On appellera $\tilde{E}_{\mathfrak{R}}(W)$ le *déroulement relatif à* $\mathfrak{R}$ *au-dessus de* W.

Si N est un espace transversal élémentaire contenu dans W et appliqué canoniquement sur $W^{\check{}}$, l'application :

$$[x, \Gamma, x'] \text{ modulo } r'_{\mathfrak{R}} \to [x^{\check{}}, \Gamma, x']_{r'(\mathfrak{R})}$$

où $x \in N$, est un isomorphisme des espaces feuilletés $\tilde{E}_{\mathfrak{R}}(N)$ sur $\tilde{E}_{\mathfrak{R}}(W)$.

Soit χ l'application :

$$[x^{\check{}}, \Gamma, x']_{r'(\mathfrak{R})} \to {}_{x^{\check{}}}\overset{\lambda}{j}_{x'}(W_1, \overset{-1}{f}, W_n)$$

de $\tilde{E}_{\mathfrak{R}}(W)$ sur $E^*(W)$, où f désigne l'isomorphisme d'holonomie associé à la chaîne pure Γ.

PROPOSITION. *Avec les notations précédentes,* χ *est un étalement de* $\tilde{E}_{\mathfrak{R}}(W)$ *sur* $E^*(W)$.

Remarquons que la topologie de $\tilde{E}_{\mathfrak{R}}(W)$ peut ne pas être séparée même si T est une topologie séparée.

En particulier, on pourra considérer pour $\mathfrak{R}$ le recouvrement formé par tous les ouverts simples; $\mathfrak{R}(F)$, où F est une feuille de (T, T'), est alors le recouvrement de F formé par toutes les plaques de F. L'espace $\tilde{E}_{\mathfrak{R}}(W)$ sera noté dans ce cas $\tilde{E}(W)$; l'application :

$$[x^{\smile}, \Gamma, x']_{r'(\mathfrak{R})} \to [x^{\smile}, \Gamma, x']_{r'}$$

est un étalement de $\tilde{E}_{\mathfrak{R}}(W)$ sur $\tilde{E}(W)$, pour tout recouvrement $\mathfrak{R}$.

Nous supposerons dans la fin de ce paragraphe que tout point de E admet un voisinage qui soit un ouvert simple dont toutes les plaques soient simplement connexes.

PROPOSITION. *Si $\mathfrak{R}'$ est un recouvrement de E formé d'ouverts simples dont toutes les plaques soient simplement connexes, toute feuille $\tilde{F}$ de $\tilde{E}_{\mathfrak{R}'}(W)$ s'identifie au revêtement universel de la feuille $\alpha^{\sim}(\tilde{F})$.*

PROPOSITION. *Soit $\mathfrak{R}'$ un recouvrement de E formé d'ouverts simples dont toutes les plaques soient simplement connexes; pour tout recouvrement $\mathfrak{R}$ de E formé d'ouverts simples, l'espace $\tilde{E}_{\mathfrak{R}'}(W)$ muni de la topologie $T_{\mathfrak{R}'}(W)$ est étalé au-dessus de l'espace $\tilde{E}_{\mathfrak{R}}(W)$ muni de la topologie $T_{\mathfrak{R}}(W)$.*

L'étalement est l'application :

$$[x^{\smile}, \Gamma, x']_{r'(\mathfrak{R}_1')} \to [x^{\smile}, \Gamma, x']_{r'(\mathfrak{R}_1)}.$$

où $\mathfrak{R}_1'$ et $\mathfrak{R}_1$ sont des recouvrements équivalents à $\mathfrak{R}'$ et $\mathfrak{R}$.

COROLLAIRE. *Si $\mathfrak{R}'$ et $\mathfrak{R}''$ sont deux recouvrements de E formés d'ouverts simples dont toutes les plaques sont simplement connexes, les espaces feuilletés $\tilde{E}_{\mathfrak{R}'}(W)$ et $\tilde{E}_{\mathfrak{R}''}(W)$ sont isomorphes.*

Nous désignerons par $\tilde{E}'(W)$ le déroulement relatif à $\mathfrak{R}'$ (ou à $\mathfrak{R}''$) au-dessus de W.

4. *Questions de stabilité*

THÉORÈME. *Soit (T, T') un feuilletage simple vérifiant les conditions suivantes :*

(1) *T' est une topologie séparée;*

(2) *Tout point $x \in E$ admet un voisinage U compact relativement à T;*

(3) *L'espace transverse $E^{\smile}$ est localement séparé.*

Si F est une feuille compacte de (T, T'), il existe un système fonda-

mental de voisinages de F saturés pour la relation d'équivalence ρ sous-jacente à (T, T') et dont toutes les feuilles sont compactes.

Démonstration. Tout voisinage de F contient un voisinage Ω de F tel que $p(\Omega)$ soit séparé, où p désigne la projection canonique de E sur $E^{\vee}$. Il existe un recouvrement $\mathfrak{U}$ fini de F par des ouverts distingués $U_i \subset \Omega$, où $i \leqslant n$, tels qu'il existe, pour tout $i \leqslant n$, un compact $U_i' \subset \Omega$ qui soit l'adhérence de U_i dans U_i' (mais pas nécessairement dans E). Soit B_i le complémentaire dans U_i' de

$$\Big(\bigcup_{i \leqslant n} U_i\Big) \cap U_i';$$

alors B_i est compact et $p(B_i)$ est un fermé ne contenant pas $x^{\vee} = p(F)$. Soit $W^{\vee}$ un voisinage de $x^{\vee}$ tel que $p(B_i) \cap W^{\vee} = \emptyset$ pour tout $i \leqslant n$. Soit

$$U = p^{-1}(W^{\vee}) \cap \Big(\bigcup_{i \leqslant n} U_i\Big).$$

Pour tout $x^{\vee\prime} \in W^{\vee}$, on a :

$$X' = \bar{p}^1(\check{x}') \cap U = \bar{p}^1(\check{x}') \cap \Big(\bigcup_{i \leqslant n} U_i'\Big) = \bigcup_{i \leqslant n} (\bar{p}^1(\check{x}') \cap U_i'):$$

donc X', étant ouvert dans $p^{-1}(x^{\vee\prime})$ et compact, est une feuille compacte contenue dans U et U est une réunion de feuilles compactes.

THÉORÈME. *Soit (T, T') un feuilletage localement simple vérifiant les conditions* (1) *et* (2) *du théorème précédent. Soit F une feuille de (T, T') compacte et fermée, dont le groupe d'holonomie est fini et telle qu'il existe $x \in F$ admettant un voisinage ouvert simple U dont l'espace transverse $U^{\vee}$ soit séparé. Alors F est propre et il existe un système fondamental de voisinages de F saturés pour la relation d'équivalence sous-jacente à (T, T') et dont toutes les feuilles sont compactes.*

En effet, soit $E^*(U)$ le déroulement de E au-dessus de U et F^* la feuille de $(T^*, T^{*\prime})$ qui se projette sur F. Alors F^* est compacte puisque le groupe d'holonomie de F est fini. Le théorème résulte du théorème précédent appliqué au feuilletage $(T^*, T^{*\prime})$.

Soit (T, T') un feuilletage localement simple sur E tel que T soit une topologie localement compacte et que tout point admette un voisinage simple dont l'espace transverse soit séparé. Alors T' est aussi localement compacte. Soit F une feuille compacte, donc propre, de (T, T').

On peut construire deux familles $(U_i)_{i \leqslant n}$ et $(U_{ij})_{(i,j) \in M}$ d'ouverts simples vérifiant les conditions suivantes :

(1) Pour tout $(i, j) \in M$, on a $i \leqslant n$, $j \leqslant n$ et M contient en particulier tout couple (i, j) tel que $U_i \cap U_j \neq \emptyset$.

(2) U_i est un ouvert distingué dans U_{ij}, pour tout $i \leqslant n$ et $(i, j) \in M$; de plus $(j, i) \in M$ et $U_{ij} = U_{ji}$.

(3) L'espace transverse $U_{ij}\check{}$ est séparé, pour tout $(i, j) \in M$.

(4) $U_i \cap F = V_i$ est connexe non vide et les V_i, où $i \leqslant n$, forment un recouvrement de F. $U_{ij} \cap F$ est connexe.

Pour tout ouvert U' saturé dans U_1, on identifiera l'espace transverse $U\check{}'$ à un sous-espace de l'espace transverse $U_1\check{}$.

Soit Δ le graphe (non orienté) ayant pour sommets les ouverts U_i, où $i \leqslant n$, et pour arêtes les ouverts U_{ij}, où $(i, j) \in M$, les extrémités de U_{ij} étant U_i et U_j. Soit A un arbre connexe de Δ contenant tous les sommets et toutes les arêtes U_{1j}.

Soient W et W' deux ouverts distingués dans un ouvert simple W''. Soit W_1'' la réunion des plaques de W'' rencontrant à la fois W et W'; si W_1 et W_1' sont les traces de W_1'' sur W et W', la suite $(W_1, W_1'', W_1'; W_1, W_1')$ est une chaîne pure. En utilisant un raisonnement par recurrence on montre la proposition suivante :

PROPOSITION. *A toute suite* $\mathfrak{U} = (U_{i_1}, \ldots, U_{i_m})$ *telle que* $(i_j, i_{j+1}) \in M$ *pour tout* $j < m$, *on associe une chaîne pure maximale*

$$C(\mathfrak{U}) = (U_1', U_{1,2}', U_2', U_{2,3}', \ldots, U_{m-1}', U_{m-1,m}', U_m'; U_1', U_2', U_2', \ldots, U_m')$$

telle que U_j' *soit saturé dans* U_{i_j} *et* $U'_{j,j+1}$ *saturé dans* $U_{i_j, i_{j+1}}$ (la relation d'ordre sur l'ensemble des chaînes pures étant celle considérée au § 3.II).

Nous désignerons par $h(\mathfrak{U})$ l'isomorphisme d'holonomie de $U_1\check{}'$ sur $U_m\check{}'$ correspondant à $C(\mathfrak{U})$.

Pour tout $i \leqslant n$, soit $\mathfrak{U}(i)$ la suite $(U_{i_1}, \ldots, U_{i_m})$ telle que $i_1 = 1$, $i_m = i$, $i_j \neq i_k$ si $j \neq k$ et $U_{i_j i_{j+1}} \in A$ pour tout $j < m$ (c'est-à-dire le chemin sur A reliant U_1 à U_i et de longueur minimale). Pour tout $(i, k) \in M$, soit $\mathfrak{U}(i, k)$ la suite $(\mathfrak{U}(i), \mathfrak{U}(k)^{-1})$

obtenue en écrivant d'abord la suite $\mathfrak{U}(i)$ puis la suite obtenue en inversant l'ordre de la suite $\mathfrak{U}(k)$.

Soit $x \in F \cap U_1$ et *supposons que le groupe d'holonomie* $\tilde{H}_x'$ *soit fini.* Pour tout $X \in \tilde{H}_x'$ choisissons une suite $\mathfrak{U}(X) = (U_1, \ldots, U_{i_m}, U_1)$ où $(1, i_2) \in M$, $(i_j, i_{j+1}) \in M$ pour tout j compris entre 2 et m, $(i_m, 1) \in M$, telle que l'isomorphisme d'holonomie correspondant à une chaîne pure contenue dans $\mathfrak{U}(X)$ ait X pour jet local en $x^\vee$, où $x^\vee$ est l'image canonique de x dans l'espace transverse $U_1^\vee$; $\mathfrak{U}(X)$ est supposée obtenue par juxtaposition d'un nombre fini de suites $\mathfrak{U}(i, k)$. Nous pouvons supposer que $\mathfrak{U}(X^{-1})$ est la suite obtenue en inversant l'ordre des termes de $\mathfrak{U}(X)$; de plus, s'il existe une suite $\mathfrak{U}(i, k)$ telle que X soit le jet local en $x^\vee$ de l'isomorphisme d'holonomie associé à une chaîne pure contenue dans $\mathfrak{U}(i, k)$, alors on choisira pour $\mathfrak{U}(X)$ une telle suite $\mathfrak{U}(i, k)$.

Nous désignerons par $\mathfrak{U}(i, k; j)$ (respectivement $\mathfrak{U}(X; j)$) la suite obtenue par juxtaposition de $\mathfrak{U}(i, k)$ (respectivement $\mathfrak{U}(X)$) et de $\mathfrak{U}(j)$. Soient $C(i, k; j)$ et $C(X; j)$ les chaînes pures correspondant à $\mathfrak{U}(i, k; j)$ et à $\mathfrak{U}(X; j)$ leur; premier élément sera désigné par $U(i, k; j)$ et $U(X; j)$ respectivement. Nous poserons : $C(X; 1) = C(X)$ et $C(i, k; 1) = C(i, k)$. Soit W l'intersection des ensembles $U(X; j)$ et $U(i, k; j)$. Soient $C'(X; j)$ et $C'(i, k; j)$ les chaînes pures contenues dans $C(X; j)$ et $C(i, k; j)$ et admettant W pour premier élément; soient $U'(X; j)$ et $U'(i, k; j)$ leurs derniers éléments.

Soit $E^*(W)$ le déroulement faible de E au-dessus de W; la feuille F^* telle que $F = \alpha^*(F^*)$ est compacte. Soient $U^{*\prime}(X; j)$ et $U^{*\prime}(i,k;j)$ les ouverts élémentaires de $E^*(W)$ définis par les chaînes pures $C'(X; j)$ et $C'(i, k; j)$. Les ouverts $U^{*\prime}(X; j)$ et $U^{*\prime}(i, k; j)$ forment un recouvrement $\mathfrak{F}$ de F^*; d'après le théorème précédent, il existe un ouvert Ω^* de $E^*(W)$ qui est une réunion de feuilles compactes et qui est contenu dans la réunion des ensembles de $\mathfrak{F}$. Posons $\omega = \beta^*(\Omega^*)$. Soient $\Gamma(i, k)$ et $\Gamma(X)$ les chaînes pures contenues dans $C(i, k)$ et $C(X)$ et admettant ω pour premier élément; leurs derniers éléments, qui sont des ouverts saturés de U_1, seront désignés par $\omega(i, k)$ et $\omega(X)$; les isomorphismes d'holonomie correspondants seront notés $h(i, k)$ et $h(X)$.

Soit ω' l'ouvert saturé de U_1 réunion des ouverts $\omega(i, k)$ et $\omega(X)$. Puisque toute la feuille passant par un point de Ω^* est contenue dans Ω^*, le pseudogroupe complété du pseudogroupe transverse d'holonomie $\Phi(\omega')$ relatif à ω' est le plus petit sous-pseudogroupe saturé par induction du pseudogroupe de tous les automorphismes locaux de l'espace transverse $\omega^{\vee\prime}$ de ω' qui contienne les isomorphismes d'holonomie $h(i, k)$ et $h(X)$. Le groupe d'holonomie $\tilde{H}_x'$ est le groupe des jets locaux en x des éléments de $\Phi(\omega')$.

THÉORÈME. *Soit B un espace topologique, O un point de B et $(f_i)_{i<n}$ une famille finie d'automorphismes locaux de B laissant O fixe. Supposons que le groupe $\mathfrak{G}$ engendré par les jets locaux $j_0^\lambda f_i$, où $i \leqslant n$, soit fini. Alors pour tout voisinage V de O il existe un ouvert W contenu dans V tel que les restrictions des applications f_i à W soient des automorphismes de W sur W qui engendrent un groupe G isomorphe à $\mathfrak{G}$, l'isomorphisme étant : $s \rightarrow j_0^\lambda s$, où $s \in G$.*

Démonstration. Soit $\mathfrak{A}(B)$ le pseudogroupe des automorphismes locaux de B; soit $\mathfrak{F}$ l'ensemble des automorphismes locaux f_i et $(f_i)^{-1}$, où $i \leqslant n$. Pour tout $\gamma \in \mathfrak{G}$, soit $g(\gamma) \in \mathfrak{A}(B)$ un pseudoproduit d'éléments de $\mathfrak{F}$ tel que $j_0^\lambda g(\gamma) = \gamma$ et que $g(\gamma^{-1}) = (g(\gamma))^{-1}$ (et par suite $g(e)$, où e est l'unité de $\mathfrak{G}$, est une unité). Soit H l'ensemble réunion de $\mathfrak{F}$ et de l'ensemble formé des $g(\gamma)$. Soit V un voisinage ouvert de O dans B; posons $V' = (\bigcap \alpha(H)) \cap V$, où $\alpha(H)$ est l'ensemble des sources $\alpha(h)$ des éléments $h \in H$. Soit H' l'ensemble des pseudoproduits hV', où $h \in H$. Pour tout triplet (i, j, k) tel que $j_0^\lambda(h_j h_i) = j_0^\lambda h_k$, où $h_i \in H'$, $h_j \in H'$ et $h_k \in H'$, il existe un ouvert $V(i, j, k)$ tel que : $h_j h_i V(i, j, k) = h_k V(i, j, k)$, $h_j h_i$ désignant le pseudoproduit. Les triplets (i, j, k) ainsi définis étant en nombre fini, posons $V'' = \bigcap V(i, j, k)$; soit H'' la classe des pseudoproduits $h'' = hV''$, où $h \in H$; on a : $h_j''h_i'' < h_k''$; posons $W = \bigcap \beta(H'')$. Pour tout $h_j'' \in H''$ et tout $h_k'' \in H''$ il existe $h_i'' \in H''$ tel que $h_j''h_i'' < h_k''$; par suite $h_j''(W) \subset h_j''(h_i''(V'') \subset h_k''(V'')$ et $h_j''(W) \subset W$ pour tout $h_j'' \in H''$. Montrons que $h_j''(W) = W$. En effet, il existe $h_{j'}'' \in H''$ tel que $h_{j'}''h_j''$ soit une unité; puisque $h_{j'}''(h_j''(W))$ est défini, on a $h_{j'}''h_j''(W) = W$; s'il existait un point $x \in W$ qui n'appartienne pas à $h_j''(W)$, il existerait un point $x' \in h_j''(W)$ tel que : $h_{j'}''(x') = h_{j'}''(x)$; puisque $h_{j'}''$ est biunivoque, il en résulterait $x' = x \in h_j''(W)$, ce qui est impossible. Donc $h''(W) = W$ pour tout $h'' \in H''$. Ceci prouve que H'' est un groupe de transformations de W.

COROLLAIRE. *Soit $\mathfrak{G}$ un groupe fini de jets locaux d'automorphismes de l'espace topologique B laissant O fixe; pour tout voisinage V de O dans B, il existe un groupe G de transformations d'un ouvert W contenu dans V laissant O fixe et tel que l'application :* $s \to j_0^\lambda s$, *où* $s \in G$, *soit un isomorphisme de G sur $\mathfrak{G}$.*

On dira alors que G est une *réalisation de $\mathfrak{G}$ par un groupe de transformations.*

Reprenons l'étude du feuilletage topologique (T, T') et du pseudogroupe transverse d'holonomie $\Phi(\omega')$ considéré plus haut. D'après le théorème précédent, il existe un voisinage ouvert $w^\vee$ de $x^\vee$ dont l'adhérence est contenue dans $W^\vee$ et tel que les restrictions à $w^\vee$ des applications $h(X)$ et $h(i, k)$ engendrent un groupe de transformations G de $w^\vee$ réalisant le groupe d'holonomie $\tilde{H}_{x'}$. Soit w l'ouvert saturé de W dont $w^\vee$ est l'espace transverse. Les chaînes pures $c(X)$ et $c(i, k)$ contenues dans $C(X)$ et $C(i, k)$ respectivement et ayant w pour premier élément sont fermées, c'est-à-dire admettent w pour dernier élément. Pour toute chaîne $C(i)$, soit $c(i)$ la chaîne pure contenue dans $C(i)$ et admettant w pour premier élément; soit w_i son dernier élément, qui est un ouvert saturé dans U_i. A une suite de plaques $\mathfrak{B} = (V_{i_1}, \ldots, V_{i_m})$ de F, où $(i_j, i_{j+1}) \in M$ pour tout $j < m$, est associée une chaîne pure notée $c(\mathfrak{B})$ ayant w_{i_1} comme premier élément et w_{i_m} comme dernier élément. Pour une chaîne de plaques pointée (x_1, γ, x_1') quelconque de F, il existe une chaîne pure pointée $(x_1, c(\mathfrak{B}), x_1')$ telle que γ et $c(\mathfrak{B})$ contiennent une même chaîne de plaques.

PROPOSITION. *Avec les notations précédentes, le groupoïde transverse d'holonomie relatif à w est le groupoïde $J^\lambda(G)$ des jets locaux des éléments de G; en particulier, pour tout $x' \in w$, $\tilde{H}_{x'}'$ est le groupe des jets locaux en $x^{\vee\prime}$ du sous-groupe $G_{x'}$ de G formé des éléments laissant $x^{\vee\prime}$ fixe.*

Par suite $\tilde{H}_{x'}'$ est isomorphe à un groupe quotient de $\tilde{H}_x'$ et il se réduit à son unité si $\tilde{H}_x'$ est réduit à son unité.

Supposons que $W^\vee$ soit localement connexe; en remplaçant, si nécessaire, $w^\vee$ par sa composante connexe contenant $x^\vee$, on peut supposer w connexe. Soit $t(w)$ la réunion des feuilles de (T, T') qui rencontrent w; alors $t(w)$ est un voisinage ouvert de F possédant les propriétés suivantes :

(1) Il existe un sous-groupoïde Γ du groupoïde des isomorphismes d'holonomie dont les unités w_i ont $t(w)$ pour réunion, où $i \in I$ et $w_1 = w$. Pour toute suite $(w_{i_1}, \ldots, w_{i_m})$, où $w_{i_{j+1}} \cap w_{i_j} \neq \emptyset$ pour tout $j < m$, il existe une chaîne pure :

$$(w_{i_1}, z_1, w_{i_2}, z_2, \ldots, z_{m-1}, w_{i_m}; w_{i_1}, w_{i_2}, \ldots, w_{i_m})$$

telle que l'isomorphisme d'holonomie associé appartienne à Γ.

(2) Il existe un ouvert simple A_i contenant l'adhérence de w_i et tel que w_i soit distingué dans A_i. L'espace transverse $w_i^{\vee}$ est considéré comme un sous-espace de l'espace transverse $A_i^{\vee}$ supposé séparé. $w^{\vee}$ est connexe et $F \cap w$ est une plaque.

(3) Pour tout $i \in I$, on a : $(t(w) \cap A_i) = w_i$.

(4) Pour tout $i \in I$, il existe un isomorphisme d'holonomie de $A_1^{\vee}$ sur $A_i^{\vee}$ qui applique $w^{\vee}$ sur $w_i^{\vee}$.

(5) Le sous-groupe de Γ ayant w pour unité s'identifie à un groupe G de transformations de $w^{\vee}$ qui est une réalisation du groupe d'holonomie de F.

(6) Toute feuille contenue dans $t(w)$ est compacte; G et I sont finis.

(6′) La frontière de $t(w)$ est identique à l'*enveloppe* de $t(w)$, c'est-à-dire à la réunion des feuilles de (T, T') qui contiennent une plaque $x^{\vee\prime}$ de A_1 appartenant à la frontière de $w^{\vee}$ dans $A_1^{\vee}$.

Remarquons que l'enveloppe de $t(w)$ est aussi la réunion des feuilles qui contiennent une plaque frontière de w_i, c'est-à-dire une plaque appartenant à la frontière de $w_i^{\vee}$ dans $A_i^{\vee}$.

Un voisinage d'une feuille F d'un feuilletage localement simple (T, T') vérifiant les conditions (1), (2), (3), (4), (6′) (respectivement (1), (2), (3), (4), (5), (6)) sera appelé un *tube* (respectivement un *tube parfait*) *d'âme* F, à groupoïde structural Γ. Tout tube parfait $t(w)$ est un tube d'âme F', pour toute feuille F' appartenant à $t(w)$. Si $t(w)$ est un tube parfait tel que le groupe G soit réduit à son unité, on appellera $t(w)$ un *tube parfait simple*; dans ce cas, $t(w)$ est aussi un tube parfait d'âme F' pour toute feuille F' appartenant à $t(w)$ et le feuilletage induit par (T, T') sur $t(w)$ est un feuilletage simple.

THÉORÈME. *Soit* (T, T') *un feuilletage localement simple tel que* T *soit une topologie localement compacte. Soit* F *une feuille compacte* *de (T, T') à groupe d'holonomie fini et telle qu'un point de F admette un voisinage simple localement connexe dont l'espace transverse soit séparé. Alors tout voisinage de F contient un tube parfait d'âme F.*

Soit $t(w)$ le tube parfait d'âme F construit ci-dessus. Nous supposerons de plus, ce qui peut toujours être réalisé, que le recouvrement $(U_i)_{i\leqslant n}$ initial vérifie de plus la condition : si $U_i \cap U_j \cap U_k \neq \emptyset$, il existe un ouvert simple U_{ijk} tel que U_{ij}, U_{jk} et U_{ki} soient des ouverts distingués dans U_{ijk}.

Soit $w^\wedge(i, s)$ l'ensemble des couples (sh_i, z), où $i \leqslant n$, $z \in w_i$, $s \in G$ et h_i est l'isomorphisme d'holonomie de $w_i^\vee$ sur $w^\vee$ associé à la chaîne pure $(c(i))^{-1}$. On munira $w^\wedge(i, s)$ de la topologie $T^\wedge(i, s)$ ayant pour base les ensembles $w^\wedge(i, s, U)$ formés des éléments (sh_i, z) tels que $z \in U$ et U est un ouvert contenu dans w_i. On munira aussi $w^\wedge(i, s)$ de la topologie $T^{\wedge\prime}(i, s)$ ayant pour base les ensembles $w^{\wedge\prime}(i, s, V)$ formés des éléments (sh_i, z) tels que $z \in V$ et V est une plaque contenue dans une plaque de w_i. Le couple $(T^\wedge(i, s), T^{\wedge\prime}(i, s))$ est un feuilletage topologique sur $w^\wedge(i, s)$.

Soit ρ la relation sur l'espace somme des espaces $w^\wedge(i, s)$ définie par : $(sh_i, z) \sim (s'h_j, z')$ si, et seulement si, $z = z'$ et $s' = ss_{ij}$, où $s_{ij} = h_i h_{ij} h_j^{-1}$ et $(h_{ij})^{-1}$ désigne l'isomorphisme d'holonomie de $w_i^\vee$ sur $w_j^\vee$ associé à la chaîne pure $(w_i, w_{ij}, w_j; w_i, w_j)$ correspondant à la suite (U_i, U_j). Si $U_i \cap U_j \cap U_k \neq \emptyset$, l'existence de U_{ijk} entraîne que l'on a :

$$h_{ji} h_{ik} = h_{jk}.$$

Il en résulte que ρ est une relation d'équivalence. Nous désignerons par $[sh_i, z]$ la classe d'équivalence de (sh_i, z) modulo ρ et par $E^\wedge(w)$ l'ensemble de ces classes modulo ρ.

Soit $T^\wedge$ (respectivement $T^{\wedge\prime}$) la topologie sur $E^\wedge(w)$ quotient de la topologie somme topologique des topologies $T^\wedge(i, s)$ (respectivement $T^{\wedge\prime}(i, s)$).

THÉORÈME. *Avec les notations précédentes, $E^\wedge(w)$ muni de la topologie $T^\wedge$ (respectivement $T^{\wedge\prime}$) est un revêtement normal de $t(w)$ muni de la topologie T (respectivement T') pour l'application $\alpha^\wedge$: $[sh_i, z] \to z$; le groupe des automorphismes de ce revêtement est isomorphe à G. Le couple $(T^\wedge, T^{\wedge\prime})$ est un feuilletage topologique simple dont l'espace transverse s'identifie à $w^\vee$ par l'application $\beta^\wedge$: $[sh_i, z] \to sh_i(z)$.*

En effet, soit $[w^\wedge(i, s)]$ l'ensemble des classes $[sh_i, z]$; les ensembles $[w^\wedge(i, s)]$, où $s \in G$, forment une partition de l'image réciproque de w_i par $\alpha^\wedge$. La restriction de $\alpha^\wedge$ à $[w^\wedge(i, s)]$ est un isomorphisme de feuilletages. A $s \in G$ correspond l'automorphisme : $[s'h_i, z] \to [ss'h_i, z]$.

$E^\wedge(w)$, muni du feuilletage $(T^\wedge, T^{\wedge\prime})$, sera appelé *déroulement parfait de* $t(w)$; l'image réciproque par $\beta^\wedge$ de $x^\vee$ est un revêtement normal $F^\wedge$ de F, que nous appellerons *âme* de $E^\wedge(w)$.

Remarque. Soit (T, T') un feuilletage topologique localement simple; soit $t(w)$ un tube d'âme F. Supposons de plus que si $w_i \cap w_j \cap w_k \neq \emptyset$ (respectivement $w_i \cap w_j \neq \emptyset$), il existe un ouvert simple A_{ijk} (respectivement A_{ij}) qui contienne w_i et A_{ij} comme distingués et détermine des éléments de Γ. Une construction analogue à la précédente permet d'obtenir un espace feuilleté $E^\wedge(w)$ qui est un revêtement normal de $t(w)$ dont le groupe des automorphismes s'identifie au sous-groupe G de Γ laissant $x^\vee$ fixe, groupe que nous appellerons *groupe structural* de $t(w)$. On appellera $E^\wedge(w)$ le *déroulement normal* de $t(w)$.

Soit $E^\wedge(w)$ le déroulement parfait du tube parfait $t(w)$ d'âme F et F' une feuille appartenant à $t(w)$. Soit w' un ouvert saturé de w et $w^{\vee\prime}$ son espace transverse qui est identifié à un sous-espace de $A_1^\vee$. Supposons que l'adhérence de $w^{\vee\prime}$ soit contenue dans $w^\vee$. Soit $x^{\vee\prime}$ une plaque de w'. On peut choisir w' de façon que $w^{\vee\prime}$ soit invariant par le sous-groupe G' de G qui laisse $x^{\vee\prime}$ fixe et que les images $s(w^{\vee\prime})$, où $s \in G$, ne rencontrent pas $w^{\vee\prime}$ pour tout $s \notin G'$. Alors la réunion $t(w')$ des feuilles rencontrant w' est un tube d'âme F' et de groupe structural G'. L'image réciproque par $\alpha^\wedge$ de $t(w')$ dans $E^\wedge(w)$ est formée de k composantes connexes, où k est le nombre de classes de G modulo G'. Une de ces composantes est l'image réciproque de $w^{\vee\prime}$ par $\beta^\wedge$ et forme un déroulement normal de $t(w')$. Nous désignerons par $F^{\wedge\prime}$ l'image réciproque de $x^{\vee\prime}$ par $\beta^\wedge$: $F^{\wedge\prime}$ est un revêtement normal de F'.

Soit $E^\wedge(w)/G'$ l'espace quotient de $E^\wedge(w)$ par la relation d'équivalence : $[sh_i, z] \sim [\sigma sh_i, z]$ pour tout $\sigma \in G'$. La classe d'équivalence $G'[sh_i, z]$ est une classe d'intransitivité de $E^\wedge(w)$ relativement à G'. Soient $T_{G'}{}^\wedge$ et $T_{G'}{}^{\wedge\prime}$ les topologies quotients de $T^\wedge$ et $T^{\wedge\prime}$ sur $E^\wedge(w)/G'$; le couple $(T_{G'}{}^\wedge, T_{G'}{}^{\wedge\prime})$ est un feuilletage topologique. Soit $\alpha_{G'}{}^\wedge$ la projection : $G'[sh_i, z] \to z$; pour cette application, $E^\wedge(w)/G'$

est un revêtement de $t(w)$. L'image de $F^{\wedge\prime}$ par l'application canonique de $E^{\wedge}(w)$ sur $E^{\wedge}(w)/G'$ est une feuille $F^{\wedge\prime}/G'$ isomorphe à F'. L'image canonique de l'âme $F^{\wedge}$ de $E^{\wedge}(w)$ est un revêtement $F^{\wedge}/G'$ de F dans lequel l'image réciproque d'un point est formée de k points.

Dans $[w^{\wedge}(i, s)]$, soit $V^{\wedge}(i, s)$ la plaque appartenant à $F^{\wedge}$ et $V^{\wedge\prime}(i, s)$ la plaque appartenant à $F^{\wedge\prime}$. Par passage au quotient, toutes les plaques $V^{\wedge}(i, \sigma s)$ (respectivement $V^{\wedge\prime}(i, \sigma s)$), où $\sigma \in G'$, sont identifiées en une plaque $V^{\wedge}(i, G's)$ (respectivement $V^{\wedge\prime}(i, G's)$) de $E^{\wedge}(w)/G'$ contenue dans $F^{\wedge}/G'$ (respectivement dans $F^{\wedge\prime}/G'$). La correspondance :

$$V^{\wedge}(i, G's) \rightarrow V^{\wedge\prime}(i, G's), \qquad \text{où } s \in G,$$

est une correspondance biunivoque. Nous désignerons par $G'[w^{\wedge}(i, s)]$ l'image canonique de $[w^{\wedge}(i, s)]$ dans $E^{\wedge}(w)/G'$.

THÉORÈME. *Avec les hypothèses du théorème page* 144 *et si tout point de F admet pour voisinage un ouvert simple dont toutes les plaques sont simplement connexes, alors il existe un tube parfait $t(w)$ d'âme F tel que le groupe fondamental de toute feuille F' contenue dans $t(w)$ soit isomorphe au groupe fondamental d'un revêtement de F.*

Démonstration. En utilisant la structure uniforme de F, on peut construire les recouvrements $(U_i)_{i \in I}$ et $(U_{ij})_{(i,j) \in M}$ et, à partir d'eux, un tube parfait $t(w)$ d'âme F de façon que la condition suivante soit réalisée:

Soit w_{ij} l'ouvert obtenu par saturation de w_i dans U_{ij} (cet ouvert est aussi obtenu par saturation de w_j dans U_{ij}). Alors il existe un recouvrement $\mathfrak{R}$ de F par des ouverts simples dont toutes les plaques sont simplement connexes et tel que, pour tout $(i, j) \in M$, la réunion des ouverts w_{kl}, où $(k, l) \in M$ et $w_{ij} \cap w_{kl} \neq \emptyset$, soit contenue dans un élément de $\mathfrak{R}$.

Les conposantes connexes des images réciproques par $\alpha_{G'}{}^{\wedge}$ dans $E^{\wedge}(w)/G'$ des éléments de $\mathfrak{R}$ forment un recouvrement $\mathfrak{R}'$ de $E^{\wedge}(w)/G'$; puisque w_{ij} est contenu dans un élément de $\mathfrak{R}$, son image réciproque par $\alpha_{G'}{}^{\wedge}$ se décompose en composantes connexes isomorphes à w_{ij}; soit $G'[w^{\wedge}(i, s; j)]$ celle de ces composantes qui contient $G'[w^{\wedge}(i, s)]$. Soient $V^{\wedge}(i, G's; j)$ et $V^{\wedge\prime}(i, G's; j)$ les plaques de $G'[w^{\wedge}(i, s; j)]$ contenues dans $F^{\wedge}/G'$ et $F^{\wedge\prime}/G'$ respectivement. La correspondance :

$$V^{\wedge}(i, G's; j) \to V^{\wedge\prime}(i, G's; j)$$

est biunivoque (puisque w_i est distingué dans w_{ij}). Pour simplifier les notations, nous poserons : $G'[w^{\wedge}(i, s; j)] = \langle i, s; j\rangle$ et $G'[w^{\wedge}(i, s)] = \langle i, s\rangle$.

Nous utiliserons la construction générale suivante : soient $(a_j)_{j \in J}$ une famille $\mathfrak{A}$ d'éléments et A' un sous-ensemble de $J \times J$ contenant la diagonale. Soit $\mathfrak{C}(\mathfrak{A})$ la catégorie dont les éléments sont les suites $(a_{j_1}, a_{j_2}, \ldots, a_{j_m})$, où $(j_\lambda, j_{\lambda+1}) \in A'$ pour tout $\lambda < m$, et dont la loi de composition est définie par :

$$(a_{j_1}, \ldots, a_{j_m})(a_{j'_1}, \ldots, a_{j'_{m'}}) = (a_{j_1}, \ldots, a_{j_m}, a_{j'_1}, \ldots a_{j'_{m'}})$$

si, et seulement si, $a_{j_m} = a_{j'_1}$. Soit $\mathfrak{C}(\mathfrak{A})/r$ le groupoïde quotient de $\mathfrak{C}(\mathfrak{A})$ par la plus petite relation d'équivalence r compatible avec la loi de composition et avec les relations élémentaires :

$$(a_j) \sim (a_j, a_j); \quad (a_j, a_k)(a_k, a_j) \sim a_j, \quad \text{où} \quad (j, k) \in A'.$$

Soit ρ_λ une relation élémentaire sur $\mathfrak{C}(\mathfrak{A})$:

$$(a_\lambda, a_{\lambda_2}, \ldots, a_{\lambda_m}, a_\lambda) \sim a_\lambda;$$

à une famille Λ de telles relations ρ_λ est associé le groupoïde $(\mathfrak{A}, A')/\Lambda$ quotient de $\mathfrak{C}(\mathfrak{A})$ par la plus petite relation d'équivalence ρ compatible avec la loi de composition et avec la relation r et les relations $\rho_\lambda \in \Lambda$; ce groupoïde est un groupoïde quotient du groupoïde libre $\mathfrak{C}(\mathfrak{A})/r$ (cette construction est un cas particulier de la construction d'un groupoïde par la donnée de générateurs et de relations, les générateurs étant ici les couples (a_j, a_k), où $(j, k) \in A'$).

Soit $\mathfrak{K}(E^{\wedge}(w)/G')$ le groupoïde $(\mathfrak{A}_w, \mathfrak{W})/\Lambda_w$ suivant : $\mathfrak{A}_w$ est la famille des ouverts $\langle i, s\rangle$, où $(i, s) \in I \times G$; $\mathfrak{W}$ est l'ensemble des couples $((i, s), (j, s'))$ tels que $(i, j) \in M$ ou $i = j$, et que $\langle j, s'\rangle$ soit contenu dans $\langle i, s; j\rangle$; Λ_w est la famille des relations :

$$(\langle i_1, s_1\rangle, \ldots, \langle i_m, s_m\rangle) \sim \langle i_1, s_1\rangle \quad \text{si} \quad \langle i_1, s_1\rangle = \langle i_m, s_m\rangle$$

et s'il existe un élément de $\mathfrak{R}'$ qui contienne $\langle i_k, s_k\rangle$, où $k \leqslant m$. Soit $\mathfrak{K}(F^{\wedge}/G')$ le groupoïde $(\mathfrak{A}_V, \mathfrak{W})/\Lambda_V$ où $\mathfrak{A}_V$ est la famille des plaques $V^{\wedge}(i, G's)$, où $(i, s) \in I \times G$, et Λ_V la famille des relations : $(V^{\wedge}(i_1, G's_1), \ldots, V^{\wedge}(i_m, G's_m)) \sim V^{\wedge}(i_1, G's_1)$ si, pour une relation appartenant à Λ_w, on a $(\langle i_1, s_1\rangle, \ldots, \langle i_m, s_m\rangle) \sim \langle i_1, s_1\rangle$. Soit de même $\mathfrak{R}(F^{\wedge\prime}/G')$ le groupoïde obtenu en remplaçant dans $\mathfrak{R}(F^{\wedge}/G')$ les plaques $V^{\wedge}(i, G's)$ par les plaques $V^{\wedge\prime}(i, G's)$.

Les groupoïdes $\mathfrak{R}(E^{\wedge}(w)/G')$, $\mathfrak{R}(F^{\wedge}/G')$, et $\mathfrak{R}(F^{\wedge\prime}/G')$ sont isomorphes. En vertu du lemme qui va suivre, il en résulte que les groupes fondamentaux de $F^{\wedge}/G'$, de $F^{\wedge\prime}/G'$ et de F' sont isomorphes.

Soit B un espace topologique dont chaque point admette un voisinage ouvert simplement connexe. Supposons donnés les éléments suivants :

(1) Un recouvrement $\mathfrak{A}$ de $\mathfrak{B}$ par des ouverts connexes a_j, où $j \in J$.

(2) Une famille d'ouverts connexes a_{jk}, où (j, k) décrit un sous-ensemble A' de $J \times J$, telle que $a_j \cup a_k \subset a_{jk}$; de plus, si $a_j \cap a_k \neq \emptyset$, on a $(j, k) \in A'$; enfin, $a_{jj} = a_j$.

(3) Un ensemble S d'ouverts simplement connexes tel que, pour tout $(j, k) \in A'$, il existe un ouvert appartenant à S qui contienne la réunion des ouverts $a_{j'k'}$ pour lesquels $a_{jk} \cap a_{j'k'} \neq \emptyset$.

Soit ρ_ν la relation sur $\mathfrak{C}(\mathfrak{A})$ (voir plus haut) : $(a_{\nu_1}, \ldots, a_{\nu_m}) \sim a_{\nu_1}$ si $\nu_1 = \nu_m$ et s'il existe un ouvert appartenant à S contenant la réunion des ouverts $a_{\nu_\mu \nu_{\mu+1}}$, où $\mu < m$. Soit Λ une famille de telles relations ρ_ν, contenant en particulier toutes les relations ρ_λ dans lesquelles on a : $a_{\lambda_\mu} \cap a_{\lambda_{\mu+1}} \neq \emptyset$ et $a_{\lambda_\mu} \cap a_{\lambda_{m-1}\lambda_1} \neq \emptyset$ pour tout $\mu < m - 1$.

Soit $\mathfrak{R}(B)$ le groupoïde $(\mathfrak{A}, A')/\Lambda$, quotient de $\mathfrak{C}(\mathfrak{A})$ par la relation d'équivalence ρ engendrée par la relation r et les relations $\rho_\nu \in \Lambda$. Les unités de $\mathfrak{R}(B)$ s'identifient aux éléments a_j, où $j \in J$.

LEMME. *Avec les hypothèses précédentes, le groupe fondamental de B est isomorphe au sous-groupe de $\mathfrak{R}(B)$ d'unité a_1.*

Démonstration. Soit x un point donné de B. Soit $\gamma = (a_{j_1}, \ldots, a_{j_m})$ un élément de $\mathfrak{C}(\mathfrak{A})$, où $x \in a_{j_1}$; désignons par $\gamma^{\wedge}$ l'ensemble des suites pointées (x, γ, x'), où x' décrit a_{j_m}; cet ensemble est muni de la topologie ayant pour base d'ouverts les ensembles $\gamma^{\wedge}(A)$ formés des suites (x, γ, x') telles que $x' \in A$ et A est un ouvert contenu dans a_{j_m}. Soit $B^{\wedge}$ l'espace quotient de l'espace somme topologique des espaces $\gamma^{\wedge}$ par la relation d'équivalence r'' : $(x, \gamma, x') \sim (x, a_{j'_1}, \ldots, a_{j'_{m'}}, x'')$ si, et seulement si, $x' = x''$ et

$(a_{j_1}, \ldots, a_{j_m}, a_{j'_{m'}}, \ldots, a_{j'_1}) \sim a_{j_1}$ modulo ρ. L'espace $B^\wedge$ est connexe: tout élément de $B^\wedge$ peut se représenter sous la forme $(a_1, \ldots, a_{j_m}, x')$, où a_1 est un ouvert choisi contenant x. Soit $\tilde{B}$ un revêtement universel de B pointé en $\tilde{x}$ au-dessus de x; soit b la projection canonique de $\tilde{B}$ sur B. La suite (x, γ, x') se relève en une suite $(\tilde{x}, \tilde{a}_{j_1}, \ldots, \tilde{a}_{j_m}, \tilde{x}')$ telle que $\tilde{a}_{j_\mu}$ et $\tilde{a}_{j_{\mu+1}}$ soient les composantes connexes de $b^{-1}(a_{j_\mu})$ et $b^{-1}(a_{j_{\mu+1}})$ qui appartiennent à une même composante connexe de $b^{-1}(a_{j_\mu j_{\mu+1}})$. L'application $\delta : (x, \gamma, x') \to \tilde{x}'$ est compatible avec la relation r''; il en résulte une application $\delta^\wedge$ de $B^\wedge$ sur $\tilde{B}$, qui définit $B^\wedge$ comme revêtement de $\tilde{B}$. Puisque $B^\wedge$ est connexe, on a $B^\wedge = \tilde{B}$. Par suite, le groupe fondamental de B s'identifie au sous-groupe de $\mathfrak{K}(B)$ formé des classes $(a_1, a_{j_1}, \ldots, a_{j_m}, a_1)$ modulo r'', c'est-à-dire au sous-groupe de $\mathfrak{K}(B)$ d'unité a_1.

Un raisonnement analogue au précédent permet plus généralement de démontrer la proposition.

PROPOSITION. *Soit (T, T') un feuilletage localement simple tel que T soit une topologie séparée. Soit F une feuille compacte telle que tout $x \in F$ admette un voisinage ouvert simple dont l'espace transverse est séparé et dont toutes les plaques sont simplement connexes. Si tout voisinage de F contient un tube $t(w)$ d'âme F, il existe un tube $t(w')$ d'âme F tel que le groupe fondamental de toute feuille F' contenue dans $t(w')$ soit isomorphe au groupe fondamental d'un revêtement de F.*

Remarque. En utilisant la structure uniforme de F, on pourrait construire les recouvrements $(U_i)_{i \in I}$ et $(U_{ij})_{(i,j) \in M}$ de F de sorte que les conditions plus strictes indiquées dans Reeb **(13)** soient vérifiées. Dans ce cas, la considération des groupoïdes $\mathfrak{K}(F^\wedge/G')$ et $\mathfrak{K}(F^{\wedge\prime}/G')$ serait remplacée par celle des nerfs des recouvrements $(V^\wedge(i, G's))$ et $(V^{\wedge\prime}(i, G's))$, où $(i, s) \in I \times G$; ces nerfs sont isomorphes. L'isomorphisme des groupes fondamentaux de F' et de $F^\wedge/G'$ résulte alors du lemme suivant, qui se démontre d'une manière analogue à celle utilisée dans le lemme précédent.

LEMME. *Soit B un espace topologique; soit $\mathfrak{A}$ un recouvrement de B par des ouverts connexes a_j, où $j \in J$, vérifiant la condition : si $a_j \cap a_k \neq \emptyset$, il existe un ouvert simplement connexe contenant la réunion de a_j et a_k. Alors le groupe fondamental de B est isomorphe au groupe fondamental du nerf du recouvrement $\mathfrak{A}$.*

THÉORÈME. *Soit* (T, T') *un feuilletage localement simple et* $t(w)$ *un tube parfait d'âme* F, *tel que l'espace transverse* $w^{\vee}$ *soit une variété topologique. Alors le groupe structural* G *de* $t(w)$ *est l'ensemble des éléments de* $\Phi(w)$ *de source* $w^{\vee}$, *où* $\Phi(w)$ *est le pseudogroupe transverse d'holonomie relatif à* w. *De plus, les déroulements* $E^*(w)$ *et* $E^{\wedge}(w)$ *sont isomorphes.*

Démonstration. Soit s l'isomorphisme d'holonomie associé à une chaîne pure fermée admettant w pour premier et dernier élément. Soit $x^{\vee\prime} \in w^{\vee}$; la feuille F' contenant $x^{\vee\prime}$ rencontre w en $p(x')$ plaques. Puisque $p(x')$ est majoré par le nombre d'éléments de G, la transformation s est périodique. Soit $s' \in G$ tel que $j_{x^{\vee}}{}^{\lambda}s = j_{x^{\vee}}{}^{\lambda}s'$; il existe un voisinage de $x^{\vee}$ dans $w^{\vee}$ dont tous les points sont fixes par l'isomorphisme d'holonomie $s^{-1}s'$. Par conséquent $s^{-1}s'$ est une transformation périodique dont l'ensemble des points fixes admet un point intérieur. En utilisant un théorème de Newman-Smith (voir **17** et **18**), on en déduit que $s^{-1}s'$ est l'identité de $w^{\vee}$, donc que $s = s' \in G$.

COROLLAIRE. *Tout élément de* $\Phi(w)$ *dont la source est connexe est une restriction d'une transformation de* G.

En effet, deux éléments distincts de G n'ayant pas de restriction commune, un agrégat de restrictions d'éléments de G dont la source est connexe est une restriction d'un élément de G.

PROPOSITION. *Avec les hypothèses du théorème précédent, l'ensemble* u *des points* $x^{\vee\prime}$ *de* $w^{\vee}$ *tels que le groupe d'holonomie* $\tilde{H}_{x'}{}'$ *soit réduit à son unité est un ouvert partout dense de* $w^{\vee}$. *Si* F' *est une feuille contenant une plaque* $x^{\vee\prime} \in u$, *alors* $t(w)$ *est un tube parfait d'âme* F'.

Démonstration. Puisque deux éléments de G qui ont le même jet local en un point sont identiques, pour tout $x' \in w$ le groupe $\tilde{H}_{x'}{}'$ est isomorphe au sous-groupe G' de G laissant fixe la plaque $x^{\vee\prime}$ qui contient x'; soit $q(x')$ le nombre d'éléments de G'. Soit $w^{\vee\prime}$ un voisinage de $x^{\vee\prime}$ ne rencontrant pas $s(w^{\vee\prime})$, pour tout $s \in G$, $s \notin G'$; pour tout $x^{\vee\prime\prime} \in w^{\vee\prime}$, on a $q(x'') \leqslant q(x')$. Si $q(x'') = q(x')$ pour tout $x^{\vee\prime\prime} \in w^{\vee\prime}$, toute transformation de G' laisse fixes tous les points de $w^{\vee\prime}$, donc est l'identité de $w^{\vee}$; par suite $q(x') = 1$. Il en résulte que, si $q(x') > 1$, il existe au moins un point $x^{\vee\prime\prime} \in w^{\vee\prime}$ tel que

$q(x'') < q(x')$. L'ensemble u des points $x^{\vee\prime}$ tels que $q(x') = 1$ est un ouvert de $w^{\vee}$; si le complémentaire de u contenait un ouvert u', il existerait $x_1^{\vee} \in u'$ tel que $q(x_1)$ soit le minimum de q sur u'; d'après ce qui précède, $q(x_1) = 1$, ce qui est contraire à l'hypothèse. Donc u est un ouvert partout dense dans $w^{\vee}$.

Remarque. Le théorème et la proposition précédents sont encore vrais dans les cas plus généraux auxquels le théorème de Smith **(18)** s'applique.

Soit (T, T') un feuilletage localement simple et F une feuille propre de (T, T').

On dira que (T, T') est *parfaitement stable autour de* F s'il existe un tube parfait d'âme F; dans ce cas, il existe aussi un système fondamental de voisinages de F formé de tubes parfaits d'âme F. En particulier, si les conditions du théorème page 144 sont vérifiées, le feuilletage (T, T') est parfaitement stable autour de toute feuille contenue dans une tube parfait.

On dira que (T, T') est *stable autour de* F s'il existe un ouvert simple U rencontrant F en une seule plaque $x^{\vee}$ et vérifiant la condition suivante : il existe des tubes $t(w)$ d'âme F tels que w soit un ouvert distingué dans U et que les images canoniques de $w^{\vee}$ dans l'espace transverse $U^{\vee}$ forment un système fondamental de voisinages de $x^{\vee}$ dans $U^{\vee}$.

THÉORÈME. *Soit (T, T') un feuilletage localement simple tel que T soit localement compact. Soit U un ouvert simple localement connexe et connexe dont l'espace transverse $U^{\vee}$ est séparé. Soit F une feuille compacte de (T, T') rencontrant U en une seule plaque $x^{\vee}$. Pour que (T, T') soit stable autour de F, il faut et il suffit qu'il existe un système fondamental de voisinages de $x^{\vee}$ dans $U^{\vee}$ qui soit invariant par le pseudogroupe transverse d'holonomie $\Phi(U')$ relatif à un ouvert U' saturé dans U.*

Démonstration. La condition étant évidemment nécessaire, montrons qu'elle est suffisante. On peut construire, comme à la page 140, deux familles $(U_i)_{i<n}$ et $(U_{ij})_{(i,j)\in M}$ ainsi que les suites $\mathfrak{U}(i, k)$ et $\mathfrak{U}(i, k; j)$. Soit U' un ouvert simple distingué dans U et dont l'adhérence est contenue dans U; soit W l'ouvert obtenu par saturation de U' dans U. On peut construire un ouvert simple W' tel qu'à toute suite $\mathfrak{U}(i, k; j)$ soit associée une chaîne pure $C(i, k; j)$ admettant W' pour premier élément; soit $U(i, k; j)$ le dernier élément

de $C(i, k; j)$. Il existe un ouvert simple connexe ω vérifiant les conditions suivantes : ω est saturé dans W'; l'adhérence du sous-espace $\omega^{\check{}}$ de l'espace transverse $U^{\check{}}$ correspondant à ω est contenue dans $W^{\check{}\prime}$; $\omega^{\check{}}$ est invariant par $\Phi(U)$. Soit $\omega(i, k; j)$ le dernier élément de la chaîne pure contenue dans $C(i, k; j)$ et admettant ω pour premier élément; l'espace transverse $\omega^{\check{}}(i, k; j)$ s'identifie à un sous-espace de l'espace transverse $U_j^{\check{}}$. Un point frontière de $\omega^{\check{}}$ est appliqué par l'isomorphisme d'holonomie correspondant à $C(i, k; j)$ sur un point frontière de $\omega^{\check{}}(i, k; j)$. Il en résulte : $\omega(i, k; 1) = \omega$ et $\omega(i, k; j) = \omega_j$ pour tout $(i, k) \in M$, $j \leqslant n$. Soit Ω la réunion des ω_j, où $j \leqslant n$. Il existe un ouvert saturé ω_j' dans ω_j dont l'adhérence est contenue dans Ω. Soit $\omega'(j)$ le premier élément de la chaîne pure contenue dans $C(1, 1; j)$ et admettant ω_j' comme dernier élément. Soit w un ouvert saturé dans U, contenu dans l'intersection des $\omega'(j)$ et tel que l'espace transverse $w^{\check{}}$ soit invariant par $\Phi(U)$. Soit w_j le dernier élément de la chaîne pure contenue dans $C(1, 1; j)$ et admettant w pour premier élément. Soit $t(w)$ la réunion des ouverts w_j, où $j \leqslant n$. Soit F' une feuille de (T, T') contenant une plaque de w; soit y un point adhérent, relativement à T', à $\Omega \cap F'$; alors y est adhérent au moins à une plaque de F' contenue dans un ouvert w_j, c'est-à-dire appartient à $t(w)$. Par suite $F' \cap t(w)$ étant ouvert et fermé dans F' est identique à F'. On en déduit que $t(w)$ est un tube d'âme F, de groupe structural le sous-groupe de $\Phi(U)$ engendré par les isomorphismes d'holonomie associés aux chaînes contenues dans $C(i, k; 1)$ et admettant w pour premier élément.

THÉORÈME. *Soit (T, T') un feuilletage topologique localement simple tel que T soit une topologie séparée et qu'il existe un tube $t(w)$ d'âme F. Si l'espace transverse $w^{\check{}}$ est une variété topologique et si toute feuille contenue dans $t(w)$ est compacte, alors (T, T') est parfaitement stable autour de F.*

Démonstration. Soit G le groupe de transformations de $w^{\check{}}$ dont les éléments sont les isomorphismes d'holonomie associés aux chaînes pures fermées appartenant au groupoïde structural Γ de $t(w)$ et ayant w pour premier et dernier élément. Puisque toute feuille F' contenue dans $t(w)$ est compacte, elle rencontre w en un nombre fini de plaques et le nombre d'éléments $p(x')$ de l'orbite de $x^{\vee\prime} \in w^{\vee}$ relativement à G est fini. Par suite, pour tout $s \in G$, il existe un nombre $k \leqslant p(x')$ tel que $s^k(x^{\vee\prime}) = x^{\vee\prime}$, c'est-à-dire s est ponctuellement périodique. D'un théorème de Montgomery (voir **19**, p. 224), il résulte que s est périodique; donc, en vertu du théorème de Newman-Smith (**17**; **18**), G est isomorphe au groupe d'holonomie $\tilde{\bar{H}}_x'$ de F par l'isomorphisme : $s \to j_{x^{\vee}}{}^{\lambda}s$. Supposons que G ne soit pas fini; alors, pour tout $x^{\vee\prime} \in w^{\vee}$, il existe $s' \in G$ différent de l'identité de $w^{\vee}$ et tel que $s'(x^{\vee\prime}) = x^{\vee\prime}$. Pour tout point $x_i^{\vee\prime}$ de l'orbite de $x^{\vee\prime}$, choisissons un élément $s_i \in G$ tel que $s_i(x^{\vee\prime}) = x_i^{\vee\prime}$. Soient u et u' deux voisinages de $x^{\vee\prime}$ dans $w^{\vee}$ tels que l'on ait : $s_i(u) \cap s_j(u) = \emptyset$ si $i \neq j$ et $s'(u') \subset u$. Pour tout $x^{\vee\prime\prime} \in u'$, on a $s_i(x^{\vee\prime\prime}) \neq s_j(x^{\vee\prime\prime})$ si $i \neq j$, d'où $p(x'') \geqslant p(x')$. Si tout $x^{\vee\prime\prime} \in u'$ était fixe pour s', la transformation périodique s' serait l'identité. Donc il existe $x^{\vee\prime\prime} \in u'$ tel que $s'(x^{\vee\prime\prime}) \neq x^{\vee\prime\prime}$; comme $s'(x^{\vee\prime\prime}) \in u$, on en déduit $p(x'') \geqslant 2p(x')$. Par récurrence, on montre que pour tout entier m, il existe $x^{\vee}{}_m'' \in u'$ tel que $p(x_m'') > m$. La fonction p étant une fonction semi-continue inférieurement sur l'espace de Baire $w^{\vee}$, il existe un ouvert sur lequel p est majorée, ce qui est contraire à l'existence de $x_m^{\vee\prime\prime}$, pour tout m. Ceci prouve que le groupe G est fini. Par conséquent $t(w)$ est un tube parfait d'âme F.

Soit (T, T') un feuilletage topologique localement simple; soit F une feuille de (T, T') et f un sous-espace de F. On appellera *chaîne pure pointée le long de* f une chaîne pure pointée (x, C, x'), où $x \in f$ et $x' \in f$, telle que la chaîne de plaques (x, c, x') correspondante sur F contienne une chaïne de connexes v_i du sous-espace f, c'est-à-dire une suite $(x, v_1, \ldots, v_m; v_{1,2}, \ldots, v_{m-1,m}, x')$ dans laquelle $v_{i,i+1}$ est une composante connexe de $v_i \cap v_{i+1}$ pour tout $i < m$. En considérant les isomorphismes d'holonomie le long de f, c'est-à-dire associés à une chaîne pure le long de f, on définit comme précédemment les notions de *groupoïde des isomorphismes d'holonomie* et de *groupe d'holonomie le long de* f.

On dira que (T, T') est *stable le long de* f si, pour toute chaîne pure (x, C, x) le long de f reliant x à x, il existe une chaîne pure fermée (x, C', x) le long de f contenue dans (x, C, x).

Supposons T localement compacte et T' séparée. Si tout point de f admet un voisinage ouvert simple dont l'espace transverse est localement connexe, connexe et séparé, si f est compacte et si le groupe d'holonomie le long de f est fini, on peut construire, à partir d'un recouvrement de f par des ouverts U_i simples, un recouvrement de f par des ouverts w_i saturés dans U_i, tel que la réunion $t(w, f)$ des ouverts w_i vérifie les conditions (1), (2), (3), (4), (5) d'un tube, dans lesquelles on remplace isomorphisme et groupe d'holonomie par isomorphisme et groupe d'holonomie le long de f.

Supposons qu'il existe un recouvrement $\mathfrak{R}$ de F par des ouverts simples dont toutes les plaques soient simplement connexes. Soit alors $\mathfrak{C}_{\mathfrak{R}}^{\sim}(x)$ le groupe fondamental* de F, identifié au groupe des classes de chaînes pures pointées $[x, C, x]$ modulo $r_{\mathfrak{R}}'$ (voir § 3). Soit $\mathfrak{C}_{\mathfrak{R}}^{\sim}(x; f)$ le sous-groupe de $\mathfrak{C}_{\mathfrak{R}}^{\sim}(x)$ formé des classes admettant un représentant (x, C, x) qui soit une chaîne pure pointée le long de f. Tout élément du groupe d'holonomie le long de f pouvant être représenté comme le jet local en $x^{\vee}$ d'un isomorphisme d'holonomie le long de f associé à une chaîne pure formée à partir du recouvrement $\mathfrak{R}$, l'application : $[x, C, x]$ modulo $r_{\mathfrak{R}}' \rightarrow j_{x^{\vee}}^{\lambda} g$, où g est l'isomorphisme d'holonomie associé à C, est un homomorphisme de $\mathfrak{C}_{\mathfrak{R}}^{\sim}(x; f)$ sur le groupe d'holonomie le long de f. Par conséquent, si $\mathfrak{C}_{\mathfrak{R}}^{\sim}(x; f)$ est fini, tout feuilletage (T, T') contenant F pour feuille est stable le long de f.

Remarquons que si $f = F$, la stabilité le long de F n'entraîne pas nécessairement la stabilité autour de F.

Rubans. Soit (T, T') un feuilletage localement simple sur E. Soit U un ouvert simple; désignons par u l'application canonique de U sur son espace transverse $U^{\vee}$. On appellera *ruban élémentaire* un sous-espace M de U qui est l'image réciproque $\bar{u}^{1}(M^{\vee})$ d'un sous-espace $M^{\vee}$ de $U^{\vee}$. Les rubans élémentaires forment une sous-espèce de structures locales de l'espèce de structures locales de tous les sous-espaces de E. Le germe $j_x^{\lambda} M$ du sous-espace M en x sera appelé *germe de ruban* en x.

L'ensemble R des germes de ruban est un ouvert de l'espace $E^{\wedge}$ de tous les germes de sous-espaces de E; sa topologie T_r a pour base les ensembles $M^{\wedge}$, où $M^{\wedge}$ est l'ensemble des germes du ruban élémentaire M. Soit T_r' la topologie sur R ayant pour base les

*Le groupe fondamental $\mathfrak{C}_{\mathfrak{R}}^{\sim}(x)$, indépendant du recouvrement $\mathfrak{R}$, s'identifie, si F est localement connexe par arcs, au groupe d'homotopie $\pi_1(F)$. Si de plus f est localement connexe par arcs, $\mathfrak{C}_{\mathfrak{R}}^{\sim}(x; f)$ s'identifié à l'image canonique de $\pi_1(f)$ dans $\pi_1(F)$.

ensembles formés des germes $j_{x'}{}^{\lambda}M$, où x' appartient à une plaque de U contenue dans le ruban élémentaire M. Le couple (T_r, T_r') est un feuilletage topologique localement simple sur R; une base d'ouverts simples est formée des ouverts $M^{\wedge}$, où M est un ruban élémentaire.

On appellera *ruban* un ensemble ouvert $\mathfrak{M}$ de R relativement à T_r, qui est une réunion de feuilles de (T_r, T_r'). Pour tout ruban élémentaire M, il existe un ruban contenant $M^{\wedge}$. Tout espace quasi-transversal détermine en chaque point un germe de ruban et il existe un ruban contenant l'ensemble de ces germes. Soit $\mathfrak{M}$ un ruban; $\mathfrak{M}$ est muni du feuilletage induit par (T_r, T_r'). La feuille $F_r^{\wedge}$ de $\mathfrak{M}$ appliquée sur la feuille F par l'application $p^{\wedge} : j_x{}^{\lambda}M \rightarrow x$ est un revêtement de F. Si $F_r^{\wedge}$ est un revêtement à k feuillets, on dira que $\mathfrak{M}$ est un ruban *multiple d'ordre k le long de F*; si $k = 1$, on dira que $\mathfrak{M}$ est *simple le long de F*. Soit $M^{\wedge}$ un ouvert simple contenu dans $\mathfrak{M}$, correspondant au ruban élémentaire M contenu dans U. Le pseudogroupe transverse d'holonomie relatif à $M^{\wedge}$ s'identifie au sous-pseudogroupe $\Phi(M)$ de $\Phi(U)$ laissant invariant le sous-espace $M^{\vee}$ de $U^{\vee}$. Les rubans simples le long de F correspondent aux germes de sous-espaces de $U^{\vee}$ invariants par le groupe des jets en $x^{\vee}$ des éléments de $\Phi(U)$ laissant $x^{\vee}$ fixe (groupe qui s'identifie au groupe d'holonomie $\tilde{H}_x')$.

On dira que (T, T') est (*parfaitement*) *stable pour F relativement au ruban* $\mathfrak{M}$ si $\mathfrak{M}$ contient une feuille $F_r^{\wedge}$ appliquée sur F par $p^{\wedge}$ et si le feuilletage de $\mathfrak{M}$ est (parfaitement) stable autour de $F_r^{\wedge}$.

Remarque sur les cas d'instabilité. Si (T, T') est un feuilletage topologique localement simple sur E et si F est une feuille compacte, à groupe fondamental infini, (T, T') est généralement instable autour de F. Si l'espace transverse $E^{\vee}$ n'est pas localement séparé, il existe des feuilles autour desquelles (T, T') est instable. Si une feuille F de (T, T') est dans l'adhérence d'une feuille F', le feuilletage est instable autour de F; il en est de même si l'adhérence du filtre des voisinages saturés (c'est-à-dire réunion de feuilles) de F ne se réduit pas à F, et par suite contient une feuille F' différente de F.

D'une manière plus précise, on peut avoir des couples de feuilles (F, F') de (T, T'), que nous appellerons *couples liés*, vérifiant les conditions suivantes :

Il existe deux voisinages simples U et U' de $x \in F$ et de $x' \in F'$ et un filtre $\mathfrak{X}$ de $\Phi(U', U)$ tels que :

(1) $\alpha(\mathfrak{X})$ est base d'un filtre convergent vers la plaque $x^{\check{}}$ et, pour tout $\xi \in \mathfrak{X}$, on a $x^{\check{}} \notin \alpha(\xi)$.

(2) La plaque $x^{\check{}\prime}$ est adhérente au filtre engendré par $\beta(\mathfrak{X})$.

Remarquons que la propriété est indépendante du choix de x et x'. Elle entraîne l'instabilité de (T, T') autour de F et de F'. Si l'espace transverse $U^{\check{}}$ est séparé, cette définition entraîne qu'il existe $\xi \in \mathfrak{X}$ tel que $x^{\check{}\prime} \notin \beta(\xi)$.

5. *Compléments relatifs à certaines structures feuilletées*

Cas de deux feuilletages supplémentaires. Soient (T, T') et (T, T_1') deux feuilletages topologiques supplémentaires sur E (c'est-à-dire sous-jacents à une structure de produit local de l'espèce $\mathfrak{T} \times \mathfrak{T}$ élargie). Alors tout point x de E admet un voisinage U qui est un ouvert simple relativement aux deux feuilletages et tel que toute plaque de U pour l'un des feuilletages soit un espace transversal élémentaire pour l'autre; U sera appelé voisinage bi-simple de x.

Supposons qu'il existe un tube $t(w)$ d'âme F relativement à (T, T') tel que les ouverts w_i soient bi-simples. Soit G le groupe structural de $t(w)$. Soit Y_i une plaque quelconque de w_i relative à (T, T_1'); à tout élément (w_j, f, w_i) du groupoïde structural Γ de $t(w)$ correspond l'homéomorphisme $f_{(Y_i, Y_j)}$ de Y_i sur Y_j appliquant $y \in Y_i$ sur $y' = Y_j \cap f(y^{\check{}})$, où $y^{\check{}}$ est la plaque de w_i relativement à (T, T') qui contient y. Soient Γ' le groupoïde de ces homéomorphismes et Γ_1' le sous-ensemble de Γ' formé de tous les homéomorphismes de la plaque Y_1 sur une plaque Y_i quelconque.

PROPOSITION. *Avec les hypothèses précédentes,* $t(w)$ *est un espace fibré* $t(w)(F, Y_1, G, \Gamma_1')$, *de base* F, *de fibre isomorphe à* Y_1 *et de groupe structural* G. *Toute feuille* F' *contenue dans* $t(w)$ *est un revêtement de* F *pour la projection de* $t(w)$ *sur* F. *Le déroulement* $E^{\wedge}(w)$ *de* $t(w)$ *est isomorphe au produit* $F^{\wedge} \times Y_1$.

Soient (T, T') et (T, T_1') deux feuilletages supplémentaires sur E; si (T, T') est stable autour de la feuille compacte F et si T est une topologie régulière, tout tube contient un tube $t(w')$ tel que les ouverts w_i' correspondants soient bi-simples; ce tube est alors muni de la structure fibrée définie ci-dessus. En particulier, si (T, T') vérifie les conditions du théorème page 144, il existe un système fondamental de voisinages de F formé de tubes parfaits $t(w)$ d'âme F qui sont des espaces fibrés; la structure fibrée de $t(w)$ est entièrement déterminée par la donnée de l'espace transverse $w^{\check{}}$.

Remarque. Si dans la proposition précédente les feuilletages (T, T') et (T, T_1') sont sous-jacents à des structures feuilletées différentiables, le groupe structural G de $t(w)$ est le groupe d'holonomie (au sens d'Elie Cartan généralisé dans (**9**)) de la connexion intégrable définie dans l'espace fibré $t(w)$ par le feuilletage (T, T'), ce qui est à l'origine de la terminologie que j'ai introduite dans l'étude des espaces feuilletées, le groupe d'holonomie $\tilde{H}_x'$ en $x \in F$ étant isomorphe au groupe $\mathfrak{J}_{x^{\check{}}}^{\lambda}(G)$ des jets locaux en $x^{\check{}}$ des éléments de G.

Réciproquement, soient (T, T') et (T, T_1') deux feuilletages supplémentaires sur E tels que (T, T') soit sous-jacent à une structure fibrée sur E de base B et de fibre isomorphe à F. Soit $(U_\iota)_{\iota \in I}$ un recouvrement de B par des ouverts connexes, l'image réciproque w_ι de U_ι dans E étant un ouvert supposé bi-simple; soit Γ le groupoïde des isomorphismes d'holonomie associés aux chaînes pures pour (T, T_1') formées à partir des ouverts w_ι. Alors E, muni du recouvrement $(w_\iota)_{\iota \in I}$, possède les propriétés (1), (3), (4) d'un tube relativement à (T, T_1'), de groupoïde structural Γ et muni d'une structure fibrée $E(B, F, G, \Gamma_1')$; la condition (2) d'un tube n'est remplie que s'il existe une feuille F_1 de (T, T_1') qui soit un relèvement de B. Une construction analogue à celle du déroulement $E^{\wedge}(w)$ d'un tube $t(w)$ permet d'obtenir un revêtement de E qui est homéomorphe au produit de F par le revêtement Γ_1' de B (voir **9**).

PROPOSITION. *Soient (T, T') et (T, T_1') deux feuilletages supplémentaires tels que E soit connexe pour T et que toutes les feuilles de (T, T') soient compactes. Si (T, T') est stable autour de toute feuille, alors une feuille F_1 de (T, T_1') rencontre toutes les feuilles de (T, T').*

Démonstration. Soit Ω le sous-espace de l'espace transverse $E^{\check{}}$ formé des feuilles de (T, T') que rencontre la feuille F_1 de (T, T_1'). Ω est ouvert. Soit F une feuille de (T, T') appartenant à l'adhérence de Ω dans E. Soit $(U_i)_{i\leqslant n}$ un recouvrement de F par des ouverts bi-simples; le voisinage $\bigcup_{i\leqslant n} U_i$ de F contenant un voisinage saturé de F, puisque F est stable, la feuille F_1 rencontre un des ouverts U_i au moins et par suite F_1 contient une plaque de cet ouvert relativement à (T, T_1'). Il en résulte que Ω contient F, donc, étant ouvert et fermé dans $E^\vee$, est identique à $E^\vee$.

La proposition est encore vraie en remplaçant l'hypothèse : (T, T') est stable autour de toute feuille F par : tout voisinage de F contient un voisinage saturé de F. Ces deux conditions sont d'ailleurs équivalentes si (T, T') est un feuilletage localement simple tel que la feuille F soit compacte et qu'un point x de F admette un voisinage simple connexe U dont l'espace transverse est séparé et localement connexe et si U est distingué dans un ouvert simple contenant l'adhérence de U.

PROPOSITION. *Avec les hypothèses de la proposition précédente, si F_1 est une variété topologique et si T est séparée, alors le déroulement faible $E^*(F_1)$ au-dessus de F_1 est un espace fibré de base F_1, de fibre isomorphe à un revêtement $F^\wedge$ de F, muni d'un feuilletage supplémentaire et admettant un revêtement isomorphe au produit de $F^\wedge$ par un revêtement de F_1.*

En effet, soit $t(w)$ un tube pour (T, T') tel que la trace Y_1 de F_1 sur w soit un espace transversal élémentaire; alors l'espace transverse $w^\vee$ est une variété topologique et, en vertu des théorèmes pages 151 et 153, $t(w)$ contient un tube parfait $t(w')$ et les déroulements $E^*(w')$ et $E^\wedge(w')$ sont isomorphes. De plus $E^\wedge(w')$ est isomorphe au produit $F^\wedge \times Y_1$, où F est un revêtement de F. La proposition résulte alors de la deuxième partie de la remarque ci-dessus.

PROPOSITION. *Soit (T, T') un feuilletage localement simple sur E tel que T soit connexe et que tout voisinage d'une feuille F de (T, T') contienne un voisinage saturé de F; alors un espace transversal fermé F_1 rencontre toute feuille.*

En effet, soit Ω l'ouvert de l'espace transverse $E^\vee$ formé des feuilles de (T, T') que rencontre F_1. Soit F une feuille appartenant à l'adhérence de Ω; si pour tout $x \in F$ il existait un voisinage $U(x)$ de x ne rencontrant pas F_1, un voisinage saturé contenu dans la réunion des ouverts $U(x)$ ne rencontrerait pas F_1, ce qui est impossible. Donc il existe un point $x' \in F$ adhérent à F_1, c'est-à-dire $x' \in F_1$, $F \in \Omega$ et $\Omega = E^{\vee}$.

Dans la démonstration de la proposition suivante, nous utiliserons le lemme :

LEMME. *Soit* (T, T') *un feuilletage simple sur* E *tel que* T *soit localement connexe et localement compact, que toute feuille soit compacte et que l'espace transverse* $E^{\vee}$ *soit localement séparé. Soit* (T, T_1') *un feuilletage localement simple sur* E; *si* F_1 *est une feuille simple de* (T, T_1') *qui est un espace transversal fermé de* (T, T'), *alors* F_1 *est un revêtement à* k *feuillets de* $E^{\vee}$, *si* $E^{\vee}$ *est connexe.*

Démonstration. La feuille F_1 étant propre, tout point x de F_1 est contenu dans un ouvert distingué relativement à (T, T') sur lequel la trace de F_1 est un espace transversal élémentaire. Par suite F_1 rencontre la feuille $x^{\vee}$ de (T, T') passant par x en k points x_i. D'après la proposition précédente, F_1 rencontre toutes les feuilles de (T, T'). Il existe un voisinage $U^{\vee}$ de $x^{\vee}$ dans $E^{\vee}$ et k relèvements s_i de $U^{\vee}$ tels que $s_i(U^{\vee}) \subset F_1$, $x_i \in s_i(U^{\vee})$ et $s_i(U^{\vee}) \cap s_j(U^{\vee}) = \emptyset$ si $i \neq j$. Soit U la réunion des feuilles de (T, T') se projetant dans $U^{\vee}$; le complémentaire F_1' dans $F_1 \cap U$ de $\cup_{i \leqslant k} s_i(U^{\vee})$ étant fermé, il existe un voisinage saturé U' de $x^{\vee}$ contenu dans U et ne rencontrant pas F_1'. Alors $F_1 \cap U'$ est une réunion d'espaces transversaux élémentaires $s_i(U^{\vee\prime}) \subset s_i(U^{\vee})$, où $i \leqslant k$ et $U^{\vee\prime}$ est l'image canonique de U' dans $U^{\vee}$. Donc F_1 est un revêtement de $E^{\vee}$.

THÉORÈME. *Soient* (T, T') *et* (T, T_1') *deux feuilletages localement simples sans point limite sur* E, *tels que* T *soit localement connexe et localement compact, que toute feuille de l'un des feuilletages soit un espace transversal pour l'autre feuilletage et que, pour chaque feuilletage, tout point de* E *admette un voisinage simple dont l'espace transverse soit localement séparé. Alors les deux feuilletages sont supplémentaires.*

Démonstration. Soit x un point de E. Il existe un voisinage U de x relativement à T vérifiant les conditions suivantes : U est un ouvert simple pour (T, T') et pour (T, T_1'), l'adhérence de U relativement à T est compacte, l'espace transverse $U^{\vee}$ relativement à (T, T') est localement séparé, l'espace transverse $U_1^\vee$ de U relativement à (T, T_1') est séparé, la plaque V_1 de (T, T_1') passant par x est un espace transversal élémentaire de (T, T'). Soient U', U'' et W des voisinages connexes de x qui sont des ouverts simples distingués dans U pour (T, T') tels que l'on ait $\bar{U}' \subset U$, $\bar{U}'' \subset U'$ et $\bar{W} \subset U''$; on peut supposer que V_1 rencontre en un point et un seul toute plaque de W ainsi que toute plaque de U'' et toute plaque de U'. Soit W' l'ouvert obtenu par saturation de W dans U'' et soit W'' la réunion des compacts qui sont l'adhérence pour T' d'une plaque de W'. Désignons par W'^* l'adhérence relativement à T du complémentaire de W' dans W''; on a $W'^* \subset U'$ et $W \cap W'^* = \emptyset$. Puisque l'adhérence de W est compact, l'adhérence de $W^\vee$ dans $U^{\vee\prime}$ est contenue dans $U^{\vee\prime\prime}$. S'il existait $y \in V_1 \cap W'^*$, on aurait $y \in \bar{U}''$, $y^\vee \in U^{\vee\prime\prime}$ et $y \notin U''$, ce qui est impossible puisque la plaque $y^\vee$ de U' rencontre V_1 en un point et un seul situé dans U''. Donc l'image canonique dans $U_1^\vee$ du compact W'^* ne contient pas V_1 et il existe un voisinage W_1 de V_1 saturé dans U relativement à (T, T_1') ne rencontrant pas un voisinage saturé de W'^*. Une plaque V_1' de W_1 a pour trace sur W'' un espace transversal pour le feuilletage induit par (T, T') sur W'', fermé dans W'' puisque trace de V_1' sur W'. D'après le lemme précédent appliqué à $(T, T')_{W''}$, V_1' rencontre toute plaque de W'' relativement à (T, T') en k points. En échangeant le rôle des deux feuilletages, on peut faire une construction analogue à partir des feuilletages induits sur $W_1 \cap W'' = W_1 \cap W'$; on obtient ainsi un voisinage w de x distingué dans W_1 pour (T, T'_1), qui rencontre la plaque V de W'' contenant x selon un espace transversal élémentaire V' de $(T, T')_{W_1 \cap W'}$ et tel que toute plaque de w relativement à (T, T') rencontre toute plaque de w relativement à (T, T_1') en un même nombre de points. Puisque V_1 rencontre toute plaque de U pour (T, T') en un point et un seul et que V' rencontre toute plaque de w relativement à (T, T_1') en un seul point, il en résulte que l'application : $y \to (y^\vee, y^\vee_1)$, où $y \in w$, est un isomorphisme de w sur $w^\vee \times w_1^\vee$, où $w^\vee$ (respectivement $w_1^\vee$) est l'espace transverse de w relativement à (T, T') (respectivement (T, T_1')), $y^\vee \in w^\vee$, $y_1^\vee \in w_1^\vee$, $y \in y^\vee \cap y_1^\vee$. Donc les deux feuilletages sont supplémentaires.

COROLLAIRE. *Soient (T, T') et (T, T_1') deux feuilletages topologiques sous-jacents à des structures de variété feuilletée sur E. Si toute feuille de (T, T_1') est un espace transversal de (T, T'), les deux feuilletages sont supplémentaires.*

En effet on construit, comme dans la proposition précédente, l'ouvert $W_1 \cap W'$. Puisque toute plaque de $W_1 \cap W'$ pour (T, T_1') est un revêtement de l'espace transverse $W^{\vee\prime}$ et que $W^{\vee\prime}$ peut être choisi simplement connexe, les composantes connexes d'une telle plaque sont isomorphes à $W^{\vee\prime}$ et $W_1 \cap W'$ est isomorphe au produit de ses espaces transverses relativement aux deux feuilletages.

Feuilletage sous-jacent à une structure de l'espèce $\mathfrak{B} \# \mathfrak{F}$ élargie. Soit (T, T') un feuilletage topologique sous-jacent à une structure Σ de l'espèce $\mathfrak{B} \# \mathfrak{F}$ élargie (les notations sont celles de § I, 4). Soit $(f, S \times S')\mathfrak{B} \# \mathfrak{F} \in \Sigma$ une structure de produit assoupli sur un ouvert simple U de (T, T') et N un espace transversal élémentaire contenu dans U. Soit ν le relèvement de $B = p(S)$ dans $B \times F$, où $F = p'(S')$, tel que $f(\nu(x)) \in N$ pour tout $x \in B$. Le germe de N au point $f\nu(x)$ s'identifie avec le jet local $j_x^{\lambda}f\nu$. Les couples $(j^{\lambda}f\nu, S)$, où $j^{\lambda}f\nu$ désigne le flot : $x \to j_x^{\lambda}f\nu$, $x \in B$, engendrent un atlas complet $A^{\wedge}$ sur l'espace transversal total $E_\theta^{\wedge}$, compatible avec $\mathfrak{B}$; ainsi $E_\theta^{\wedge}$ est muni d'une structure de l'espèce $\mathfrak{B}$ élargie. Cette structure ne dépend que de la structure feuilletée sous-jacente de l'espèce $\mathfrak{B} \# \mathfrak{T}$ élargie; elle est entièrement déterminée par la structure feuilletée de seconde espèce $(\mathfrak{B}, \mathfrak{H}^{-} (\mathfrak{B} \# \mathfrak{F}))$ sous-jacente à Σ. Elle induit sur chaque espace transversal une structure de l'espèce $\mathfrak{B}$ élargie.

Toute chaîne pure contient une chaîne pure C formée d'ouverts simples munis d'une structure de produit assoupli. Les espaces transverses $U_1^{\vee}$ et $U_n^{\vee}$ du premier et du dernier élément de C sont munis de structures $(f_1, S_1)\mathfrak{B}$ et $(f_n, S_n)\mathfrak{B}$ de l'espèce $\mathfrak{B}$ élargie. Si (U_n, γ, U_1) désigne l'isomorphisme d'holonomie associé à C, alors $((f_n, S_n)\mathfrak{B}, \gamma, (f_1, S_1)\mathfrak{B})$ appartient au groupoïde $\mathfrak{B}^{-}$ des isomorphismes de l'espèce $\mathfrak{B}$-élargie. On définit ainsi un isomorphisme du groupoïde des isomorphismes d'holonomie H' de (T, T') sur un sous-groupoïde $H'(\Sigma)$ de $\mathfrak{B}^{-}$ qui sera appelé *groupoïde des isomorphismes d'holonomie de* Σ. Cet isomorphisme détermine aussi un isomorphisme du groupoïde d'holonomie de (T ,T') sur un sous-groupoïde $\tilde{H}'(\Sigma)$ du groupoïde des jets locaux de $H'(\Sigma)$, que nous appellerons *groupoïde d'holonomie de* Σ. Le sous-groupe de $\tilde{H}'(\Sigma)$ correspondant à un point x d'une feuille sera appelé *groupe d'holonomie de* Σ *en* x. De même au pseudogroupe transverse d'holonomie relatif à U correspond un pseudogroupe transverse d'holonomie de Σ relatif à U, qui est un pseudogroupe de transformations de l'espace transverse $U^{\check{}}$ muni de sa structure de l'espèce $\mathfrak{B}$-élargie. Les feuilletages $(T^*, T^{*\prime})$ et $(\tilde{T}_{\mathfrak{R}}, \tilde{T}_{\mathfrak{R}}')$ des déroulements $E^*(U)$ et $\tilde{E}_{\mathfrak{R}}(U)$ au-dessus de U sont sous-jacents à des structures feuilletées de l'espèce $\mathfrak{B} \# \mathfrak{F}$ élargie. Si $t(U)$ est un tube d'âme F', le déroulement $E^{\hat{}}(U)$ de $t(U)$ est aussi muni d'une structure feuilletée de l'espèce $\mathfrak{B} \# \mathfrak{F}$ élargie, dont le feuilletage topologique sous-jacent est le feuilletage $(T^{\hat{}}, T^{\hat{}\prime})$.

THÉORÈME. *Soit* (T, T') *un feuilletage topologique sous-jacent à une structure feuilletée de l'espèce* $\mathfrak{B} \# \mathfrak{F}$ *élargie, tel que* T' *soit une topologie localement connexe et localement compacte. Alors* (T, T') *est parfaitement stable autour de toute feuille compacte, à groupe d'holonomie fini.*

La construction utilisée au § 4 se modifie aisément, les ouverts simples considérés étant tous supposés munis d'une structure de produit assoupli.

Soit $\Sigma \in \mathfrak{L}(\mathfrak{B} \# \mathfrak{F})_0$ une structure feuilletée sur E, où $\mathfrak{B}$ est le groupoïde des isomorphismes r fois différentiables d'une variété sur une autre. On peut définir dans ${}_{x'}\mathfrak{T}'_x$ (page 132) la relation d'équivalence ${}_{x'}\rho_x{}^r$ suivante : $(U', f, U) \sim (U_1', f_1, U_1)$ si, et seulement si, il existe un ouvert U_2 distingué dans U et dans U_1 et un ouvert U_2' distingué dans U' et dans U_1' tels que l'on ait : ${}_{x^{\check{}\prime}}j_{x^{\check{}}}{}^r u'^{-1} f u = {}_{x^{\check{}\prime}}j_{x^{\check{}}}{}^r u_1'^{-1} f_1 u_1$, où $x^{\check{}}$ et $x^{\check{}\prime}$ sont les plaques de U_2 et U_2' contenant x et x', u, u_1 (respectivement u', u_1') les applications canoniques de l'espace transverse $U^{\check{}}_2$ sur $U^{\check{}}$ et $U_1{}^{\check{}}$ (respectivement $U_2{}^{\check{}\prime}$ sur $U^{\check{}\prime}$ et $U_1{}^{\check{}\prime}$).

La classe d'équivalence de (U', f, U) modulo ${}_{x'}\rho_x{}^r$ sera désignée par ${}_{x'}j_x{}^r(U', f, U)$ et appelée *jet transverse d'ordre* r, de source x et de but x'. L'ensemble des jets transverses d'ordre r forme un groupoïde $J^r(\mathfrak{J}')$. Le sous-groupoïde $J^r(H')$ de $J^r(\mathfrak{J}')$ formé des jets transverses d'ordre r ${}_{x'}j_x{}^r(U', f, U)$, où $(U', f, U) \in H'$, sera appelé *groupoïde d'holonomie d'ordre* r et noté $\tilde{H}'^{r}$; un élément de $\tilde{H}'^{r}$ est appelé *jet d'holonomie d'ordre* r, le sous-groupe de $\tilde{H}'^{r}$ formé des jets de source et but x, groupe d'holonomie d'ordre r en x.

Soient U et U' deux ouverts simples munis de structures de produits assouplis. En identifiant un jet transverse d'ordre r, ${}_{x'}j_x^{r}(U_1', f, U_1)$, où U_1 et U_1' sont distingués dans U et U' respectivement au triplet $(x', {}_{x^{\smile\prime}}j_{x^{\smile}}^{r}f, x)$, l'ensemble de ces jets est identifié à un sous-espace de l'espace topologique $U' \times J^{r}(U^{\smile\prime}, U^{\smile}) \times U$ qui sera noté $J^{\smile r}(U', U)$; le sous-ensemble de $J^{\smile r}(U', U)$ formé des jets de source x et de but x' sera noté $J^{\smile r}(x', x)$ et s'identifie à ${}_{x^{\smile\prime}}J_{x^{\smile}}^{r}(U^{\smile\prime}, U^{\smile})$. Soit $J^{\smile r}(V', V)$ le sous-espace de $J^{\smile r}(U', U)$ formé des jets transverses dont la source décrit V et le but V', où V et V' sont deux plaques de U et U' respectivement. Soit T^{r} (respectivement T'^{r}) la topologie sur $J^{r}(\mathfrak{F}')$ ayant pour base les ouverts des espaces topologiques $J^{\smile r}(U', U)$ (respectivement $J^{\smile r}(V', V)$).

PROPOSITION. *Muni de la topologie T^{r} (respectivement T'^{r}) définie ci-dessus, $J^{r}(\mathfrak{F}')$ est un groupoïde topologique localement trivial* **(7)**. *Le couple (T^{r}, T'^{r}) est un feuilletage topologique sous-jacent à une structure feuilletée de l'espèce $(\mathfrak{B} \times \mathfrak{T} \times \mathfrak{B}) \# (\mathfrak{F} \times \mathfrak{T} \times \mathfrak{F})$ élargie.*

Nous supposerons désormais que Σ est une structure feuilletée d'espèce $\Lambda_p^{r} \# \mathfrak{F}$ élargie. Alors $J^{r}(\mathfrak{F}')$ muni de la topologie T^{r} (respectivement T'^{r}) est un espace fibré de base $E \times E$, dont les fibres $J^{\smile r}(x', x)$ sont isomorphes à L_p^{r}, le groupe structural étant $L_p^{r} \times L_p^{r}$, où L_p^{r} désigne le groupe des jets inversibles d'ordre r de source et de but l'origine de l'espace numérique $R.^{p}$

Un espace topologique E' admettant $J^{r}(\mathfrak{F}')$ muni de la topologie T^{r} comme groupoïde d'opérateurs topologique est appelé un *prolongement transverse* de Σ; il admet aussi $\tilde{H}'^{r}$ pour groupoïde d'opérateurs. Chaque prolongement transverse de Σ est un espace fibré de base E et de groupe structural L_p^{r}. Réciproquement si K est un espace topologique admettant L_p^{r} pour groupe d'opérateurs topologique, par élargissement **(3)** de K (considéré comme espèce de structures sur L_p^{r}) au-dessus de $J^{r}(\mathfrak{F}')$, on obtient un prolongement transverse de Σ.

Soit E' un prolongement transverse de Σ obtenu par élargissement de K; soient F une feuille de Σ passant par x et q l'application canonique de E' sur E. Le sous-espace $F' = \bar{q}^1(F)$ est un espace fibré de base F, de groupe structural le groupe d'holonomie $\tilde{H}_x'^r$ d'ordre r en x, muni de la topologie discrète; ainsi F' admet une connexion intégrable. F' sera appelé *prolongement transverse de F dans E'*. Soit $J_x^r(\mathfrak{F}')$ le sous-espace de $J^r(\mathfrak{F}')$ muni de la topologie T^r formé des jets transverses d'ordre r de source x; soit W un ouvert simple de E et $x' \to h_{x'}$ un relèvement continu de W dans $J_x^r(\mathfrak{F}')$; soit η l'application : $(x', y) \to h_{x'}.y$, où $x' \in W$, $y \in \bar{q}^1(x)$ et où $h_{x'}.y$ désigne le composé de $h_{x'}$ avec y. Les applications η ainsi définies engendrent un atlas complet qui détermine sur E' une structure fibrée, de topologie sous-jacente t. En remplaçant W par une plaque V de W et la topologie T^r par T'^r, on obtient de même une deuxième structure fibrée sur E', de topologie t'. Le couple (t, t') définit un feuilletage topologique sur E' dont les feuilles sont les prolongements transverses dans E' des feuilles de Σ.

PROPOSITION. *Avec les hypothèses précédentes, si les feuilles F de Σ, munies de leur seule structure topologique, sont les fibres d'une fibration localement triviale ayant pour base un espace localement contractile, alors les feuilles F' de (t, t') sont aussi les fibres d'une fibration localement triviale de même base.*

La démonstration utilise le théorème 1 de **(8)**.

Un exemple de prolongement transverse de Σ est obtenu de la façon suivante : soit M une variété r fois différentiable. Dans l'ensemble des couples (g, U), où U est un ouvert simple de E et g une application r fois différentiable de $U^{\check{}}$ dans M, soit ρ_x la relation d'équivalence définie par : $(g, U) \sim (g', U')$ si, et seulement si, il existe un ouvert U'' distingué dans U et dans U' tel que l'on ait : $j_x^r gu = j_{x^{\check{}}}^r g'u'$ où $x \in U$, u et u' sont les applications canoniques de l'espace transverse $U^{\check{}\prime\prime}$ dans $U^{\check{}}$ et $U^{\check{}\prime}$, $x^{\check{}}$ est la plaque de U'' contenant x.

La classe d'équivalence modulo ρ_x de (g, U) sera appelée *jet transverse d'ordre r de E dans M, de source x et de but* $g(x)$. En particulier, un jet transverse d'ordre r de E dans la droite numérique, de but O, est appelé *covitesse transverse d'ordre r*. L'ensemble des covitesses transverses d'ordre r de source x forme un espace vectoriel K_x^r isomorphe à $L_{l,p}^{r*}$. L'ensemble de toutes les covitesses d'ordre r sur E est un prolongement transverse de Σ. En remplaçant

K_x^r par un espace tensoriel au-dessus de K_x^r, on obtient un espace de *tenseurs transverses d'ordre r* en x. L'ensemble de tous les tenseurs transverses d'ordre r d'un certain type est aussi un prolongement transverse de Σ.

Supposons que Σ est une structure de variété feuilletée sur E d'ordre (k, l, r). Alors l'espace transversal total est muni d'une structure différentiable d'ordre l; muni de sa structure différentiable, un espace transversal sera appelé *variété transversale.* Le pseudogroupe d'holonomie H (voir page 130) s'identifie à un pseudogroupe d'homéomorphismes l fois différentiables d'une variété transversale élémentaire sur une autre. Soit $H^{\wedge l}$ (respectivement $H_x^{\wedge l}$) le groupoïde des jets d'ordre l des éléments de $H^\wedge$ (respectivement $H_x^\wedge$). Le groupoïde $\tilde{H}^l$ quotient de $H^{\wedge l}$ par le groupoïde réunion des groupoïdes $H_x^{\wedge l}$, où $x \in E$, s'identifie au groupoïde d'holonomie d'ordre l de E; ses unités sont les groupoïdes $H_x^{\wedge l}$ appelés *germes transverses d'ordre l* en x. L'espace en x des vecteurs transverses d'ordre 1 s'identifie à l'espace quotient de l'espace des vecteurs tangents en x à E par le sous-espace des vecteurs tangents en x à la feuille F passant par x.

PROPOSITION. *Pour qu'une sous-variété propre F d'une variété différentiable soit une feuille d'une structure feuilletée définie dans un voisinage de F, il suffit que l'espace des vecteurs transverses admette une connexion intégrable* (voir **9**).

Etant donnée une variété k fois différentiable V_p, rappelons qu'une G-structure d'ordre s et de classe s' sur V_p est une section s' fois différentiable au-dessus de V_p de l'espace quotient par G du prolongement principal $H^s(V_p)$, où G est un sous-groupe fermé du groupe L_p^s et $s' \leqslant k - s$. Un prolongement d'ordre s d'une G-structure d'ordre s et de classe s' est un espace fibré s' fois différentiable associée au sous-espace fibré de l'espace fibré $H^s(V_p)$ formé des repères distingués de la G-structure. Soit $\mathfrak{F}_0$ l'espèce des G-structures d'ordre s et de classe s' sur les variétés k fois différentiables de dimension p. Soit $\mathfrak{F}_0'$ l'espèce de structures formée par les prolongements d'ordre s des éléments de $\mathfrak{F}_0$ correspondant à une représentation s' fois différentiable de G sur un groupe d'automorphismes s' fois différentiables d'une variété K. Soient $\mathfrak{F}$ et $\mathfrak{F}'$ les groupoïdes des isomorphismes correspondant à $\mathfrak{F}_0$ et $\mathfrak{F}_0'$. Soit $\Sigma \in \mathfrak{L}(\mathfrak{B} \# \mathfrak{F})_0$ une structure feuilletée sur E. A une carte

$(f, S \times S') \in \Sigma$ correspond la carte $(f', S \times S'')$, où $S'' \in \mathfrak{F}_0'$ est le prolongement de S' et où l'application $f_x' : Y \to f'(x, Y)$ est le prolongement de $f_x : y \to f(x, y)$ à S'' (voir **10**). Les cartes $(f', S \times S'')$ engendrent un atlas complet Σ' sur un ensemble E' compatible avec $\mathfrak{B} \# \mathfrak{F}'$. Muni de la structure feuilletée Σ' d'espèce $\mathfrak{B} \# \mathfrak{F}'$ élargie, E' sera appelé *prolongement de* Σ *relativement à* $\mathfrak{F}'$; les feuilles de Σ' sont des prolongements des feuilles de Σ. L'atlas Σ' définit aussi sur E' une structure d'espace fibré de base E et de fibres isomorphes à K.

PROPOSITION. *Avec les hypothèses précédentes, si les feuilles de* Σ, *munies de leur seule structure topologique, sont les fibres d'une fibration localement triviale dont l'espace quotient est un espace localement contractile, alors les feuilles de* Σ' *sont aussi les fibres d'une fibration localement triviale de même base.*

Bien que les prolongements des feuilles de Σ soient isomorphes en tant qu'espaces fibrés, les G-structures sur ces feuilles ne sont pas nécessairement isomorphes.

Structures feuilletées à métrique riemannienne transverse. Si $\mathfrak{B}$ est le groupoïde des isomorphismes trois fois différentiables d'une structure riemannienne trois fois différentiable sur une autre, une structure feuilletée Σ sur E d'espèce $\mathfrak{B} \# \mathfrak{F}$ élargie est appelée *structure feuilletée à métrique riemannienne* transverse. Tout espace transversal de Σ dont la topologie est séparée est alors muni d'une métrique riemannienne. Pour qu'une structure feuilletée sur E appartienne à $\mathfrak{L}(\mathfrak{B} \# \mathfrak{F})$, il suffit qu'il existe un espace transversal $\mathfrak{N}$ tel que E soit la réunion des feuilles rencontrant $\mathfrak{N}$ et que $\mathfrak{N}$ admette une métrique riemannienne invariante par le pseudogroupe d'holonomie relatif à $\mathfrak{N}$ (c'est-à-dire le sous-pseudogroupe du pseudogroupe d'holonomie dont les unités sont contenues dans $\mathfrak{N}$).

PROPOSITION. *Soit* Σ *une structure feuilletée sur* E *à métrique riemannienne transverse; s'il existe un tube* $t(w)$ *d'âme* F, *le feuilletage topologique sous-jacent à* Σ *est stable autour de* F.

Dans ce cas, le déroulement normal $E\hat{}(w)$ de $t(w)$ est isomorphe au déroulement faible $E^*(w)$ au-dessus de w et il existe un tube d'âme F dont le groupe structural est isomorphe à un groupe de rotations.

Soit (T, T') un feuilletage topologique sur E sous-jacent à une structure feuilletée Σ à métrique riemannienne transverse; supposons T localement compact. Soit (T, T_1') un feuilletage topologique supplémentaire à (T, T').

THÉORÈME. *Avec les hypothèses précédentes et si toute feuille F_1 de (T, T_1') est munie d'une structure riemannienne complète, toute feuille F de Σ admet un voisinage saturé qui est une réunion d'ouverts bi-simples et (T, T') est stable autour de toute feuille propre. F_1 rencontre toute feuille F et le déroulement $E^*(F_1)$ au-dessus de F_1 est un espace fibré de base F_1 de fibres isomorphes à un revêtement $F^\wedge$ de F, admettant un revêtement isomorphe à $F^\wedge \times F_1^\wedge$, où $F_1^\wedge$ est un revêtement de F_1.*

Démonstration. Soit F une feuille de Σ. Pour tout $x \in F$, désignons par $B(x, r)$ la boule fermée de centre x de rayon r contenue dans la feuille F_x de (T, T') contenant x. Soit r_x le plus grand nombre tel que tout point intérieur à $B(x, r_x)$ puisse être joint à x par un arc de géodésique unique. Soit $t(w_x, B(x, r_x))$ un des ensembles analogues à un tube considérés page 155 recouvrant $B(x, r_x)$, réunion d'une famille ϕ d'ouverts bi-simples, $w_x \in \phi$ contenant x; on montre comme au début du paragraphe que $t(w_x, B(x, r_x))$ est un ouvert bi-simple; soit $W(x, r)$ l'ouvert bi-simple contenu dans l'ouvert $t(w_x, B(x, r_x))$, produit de la plaque $x^\vee$ de w_x pour (T, T') contenant x par l'intérieur de la boule $B(x, r)$. Pour tout $x' \in x^\vee$, on a $r_{x'} \geqslant r_x$. Soit V l'ouvert de F, supposé non vide, formé des points x tels que $r_x > r$, où r est un nombre fixé. Soit x un point de F adhérent à V; montrons que x appartient à V. Pour tout $x' \in V \cap x^\vee$, désignons par $p_{x'}$ l'homéomorphisme de $B(x', r_x)$ sur $B(x, r_x)$ déterminé par la projection canonique de $t(w_x, B(x, r_x))$ sur sa plaque pour (T, T_1') contenant x. Cet homéomorphisme se prolonge en une application continue $p_{x'}'$ de $B(x', r)$ dans F_x en associant à l'extrémité z d'un arc de géodésique (x', z) de longueur $r' \leqslant r$ le point z' situé à la distance r' de x sur l'arc de géodésique contenant $p_{x'}(x', z)$. Soit $g_{x'}$ un arc de géodésique issu de x' de longueur r; soit $[g_{x'}]$ le plus grand nombre tel que, pour tout $z \in g_{x'}$ situé à une distance de x' inférieure à $[g_{x'}]$, la restriction de $p_{x'}'$ à un voisinage assez petit de z soit un homéomorphisme sur son image; si $x'' \in V \cap x^\vee$ et si $p_{x''}'(g_{x''}) = p_{x'}'(g_{x'}) = g_x$, on a $[g_{x''}] = [g_{x'}] = [g]$. Soient $(x, \xi_{x'})$ et (x, η) les arcs de $g_{x'}$ et g_x de longueur $[g]$. On peut construire une chaîne pure pour (T, T_1') reliant x à η, formée d'ouverts bi-simples U_i, $i \leqslant n$, recouvrant (x, η); pour tout point $x'' \in x^{\vee} \cap V$ appartenant à la plaque de U_1 pour (T, T') contenant x, U_n détermine un homéomorphisme d'un voisinage de $\xi_{x''}$ dans $F_{x''}$ sur un voisinage de η. Il en résulte que $[g] = r$ et que, pour tout $x' \in V \cap x^{\vee}$, l'application $p_{x'}'$ est un étalement de $B(x', r)$ dans F_x; $B(x', r)$ étant compact, $B(x', r)$ est un revêtement de F_x. Puisque l'image réciproque de x par $p_{x'}'$ est réduite à x', $p_{x'}'$ est un isomorphisme de $B(x', r)$ sur $B(x, r)$. Ainsi $x \in V$ et $V = F$. Montrons que l'ouvert $t(W(x, r))$ réunion des ouverts bi-simples $W(x', r)$, où $x' \in F$, est saturé. Soit F' une feuille de (T, T') rencontrant $t(W(x, r))$; soit $V' = F' \cap t(W(x, r))$; V' est ouvert dans F'. Soit z un point adhérent à V' dans F'; alors $z \in F'$. En effet, soit u un ouvert bi-simple contenant z; soit g_z un arc de géodésique dans F_z issu de z; pour tout point z' appartenant à la plaque de u pour (T, T') contenant z, soit $g_{z'}$ un arc de géodésique issu de z' et tel que $g_{z'} \cap u$ soit appliqué sur $g_z \cap u$ par l'application canonique de u sur la plaque de u pour (T, T_1') contenant z; soient $(z, \bar{z})$ et $(z', \bar{z}')$ les arcs de g_z et $g_{z'}$ de longueur k. Il existe une chaîne pure pour (T, T_1') reliant z à $\bar{z}$, formée d'ouverts bi-simples u_i, où $i < n$ et $u_1 \subset u$, recouvrant $(z, \bar{z})$ et telle que $u_{i-1} \cap u_i$ contienne un ouvert bi-simple u_i' admettant u_{i-1} et u_i pour saturés dans u_{i-1} et u_i respectivement. Désignons par $z_1^{\vee}$ et $z_n^{\vee}$ les plaques de u_1 et u_{n-1} pour (T, T') contenant z et $\bar{z}$ respectivement, par $z_i^{\vee}$ une plaque de u_i' pour (T, T') contenant un point z_i choisi sur $(z, \bar{z})$. Pour tout $z' \in z_1^{\vee}$, posons $z_i' = g_{z'} \cap z_i^{\vee}$; les arcs (z_i', z_{i+1}') et (z_i, z_{i+1}) de $g_{z'}$ et g_z respectivement ont même longueur et l'application : $z_i' \to z_{i+1}'$ est continue. On en déduit que $z_n' = \bar{z}'$ et que l'application $\gamma_k : z' \to \bar{z}'$, où $z' \in z_1^{\vee}$, est une application continue de $z_1^{\vee}$ sur $z_n^{\vee}$. En particulier, pour $k = r$, on a $\gamma_r(z') \in F$ si $z' \in V'$, d'où $\gamma_r(z) \in F$ et $z \in t(W(x, r))$. Donc $t(W(x, r))$ est saturé. Le reste du théorème se démontre facilement.

COROLLAIRE (THÉORÈME DE DE RHAM (**20**)). *Un espace riemannien complet réductible et simplement connexe est isomorphe au produit de deux espaces riemanniens.*

Remarques.

(1) Une démonstration analogue à la fin de celle du théorème permet de prouver le résultat plus général suivant : soit v_z un

vecteur tangent en z à la feuille F_z de (T, T_1'); soit $\bar{z}_l$ l'extrémité de l'arc de géodésique issu de z tangent en z à v_z et de longueur l; alors l'application : $(v_z, l) \rightarrow \bar{z}_l$ est une application continue pour les topologies du prolongement des vecteurs transverses, de la droite numérique et pour la topologie T.

(2) Si (T, T') et (T, T_1') sont deux feuilletages topologiques sur E supplémentaires tels que, pour toute feuille de (T, T'), il existe un tube d'âme F formé d'ouverts bi-simples, alors toute feuille de (T, T_1') rencontre toute feuille de (T, T').

(3) En prenant pour $\mathfrak{B}$ le groupoïde des isomorphismes d'espaces localement homogènes (de Lie **(11)**) ou le groupoïde des isomorphismes 3 fois différentiables d'une variété munie d'une connexion de Cartan **(9)** 2 fois différentiable sur une autre, une structure feuilletée d'espèce $\mathfrak{B} \# \mathfrak{F}$ élargie sera dite *à structure localement homogène transverse*, ou *à connexion de Cartan transverse. Pour une structure feuilletée Σ sur E à structure localement homogène transverse ou à connexion de Cartan transverse, le théorème précédent est valable en remplaçant structure riemannienne complète par structure localement homogène complète ou connexion de Cartan complète.*

(4) Soit Σ une structure feuilletée d'espèce $\Lambda_p{}^r \# \mathfrak{F}$ élargie. Soit $\mathfrak{B}_0$ l'espèce de structures locales formée des variétés r fois différentiables munies d'une forme extérieure; soit $\mathfrak{B}$ le groupoïde des isomorphismes correspondant. Si Σ admet une superstructure $\Omega \in \mathfrak{L}(\mathfrak{B} \# \mathfrak{F})_0$, on appellera Ω une *forme extérieure transverse* sur Σ. L'application covariante d de différentiation extérieure s'élargit en une application associant à Ω une forme extérieure transverse $d\Omega$ sur Σ appelée *différentielle extérieure de* Ω. L'ensemble des formes extérieures transverses sur Σ est un module différentiel gradué, d'où on déduit le *module d'homologie transverse.*

Remarque sur la notion de simultanéité. Le feuilletage topologique sous-jacent à une structure de variété feuilletée n'admet généralement pas de feuilletage supplémentaire. Ainsi un espace fibré simplement connexe n'admet un feuilletage supplémentaire à la fibration que s'il est isomorphe au produit de sa base par une fibre, les fibres étant supposées compactes. Par exemple la sphère S_3 n'admet aucun feuilletage supplémentaire à sa fibration classique.

Soit V_4 un univers de la théorie de la relativité. La notion de simultanéité dans V_4 sera définie par un feuilletage topologique dont les feuilles F, appelées *feuilles de simultanéité*, sont de caractère spatial en chaque point x (c'est-à-dire la restriction du ds^2 fondamental à l'espace tangent à F en x est une forme quadratique définie positive). L'exemple de l'espace fibré $S_3 \times S_1$ de base $S_2 \times S_1$ montre qu'un tel feuilletage de simultanéité n'existe généralement pas.

Si l'univers V_4 admet un feuilletage de simultanéité, il existe un feuilletage topologique supplémentaire dont les feuilles sont des *lignes de temps* (c'est-à-dire telles que la restriction du ds^2 à la tangente en chaque point est définie négative). Une feuille de simultanéité peut rencontrer une ligne de temps en plusieurs points.

Rappelons le résultat suivant **(12)** :

PROPOSITION. *Une variété compacte V_4 admet un ds^2 admissible en relativité si, et seulement si, la caractéristique d'Euler-Poincaré de V_4 est nulle.* (En considérant un revêtement à deux feuillets de V_4, on se ramène au cas considéré dans **(12)** où l'on peut définir continuement en chaque point un demi-cône de lumière, déterminant l'orientation vers l'avenir.)

PROPOSITION. *Un univers compact V_4 admet un feuilletage dont les feuilles sont les lignes de temps et dont une feuille est fermée. Si V_4 admet aussi un feuilletage de simultanéité, alors toute ligne de temps rencontre toute feuille de simultanéité.*

Soit V_4 un univers muni d'un feuilletage de simultanéité donné; pour qu'on puisse définir localement sur V_4 une *durée*, il faut et il suffit que ce feuilletage soit à métrique riemannienne transverse. S'il en est ainsi et si de plus les lignes de temps ont pu être munies continuement d'une orientation (passé-avenir), la durée est définie par la donnée d'une forme transverse fermée ω de degré 1. Pour tout arc γ transversalement différentiable, on peut définir $\int_\gamma \omega$ qui représentera la durée correspondant à γ. En appliquant le théorème précédent, on obtient en particulier la proposition suivante (qui résulte aussi de l'étude faite par Reeb d'un feuilletage défini par une forme fermée **(13**, p. 110**)**.

PROPOSITION. *Avec les hypothèses précédentes, si V_4 est compact toutes les feuilles de simultanéité sont isomorphes; l'une de ces feuilles* *est soit partout dense, soit compacte; dans ce dernier cas,* V_4 *est un espace fibré de base* S_1 *et de fibre isomorphe à une feuille de simultanéité. Si* V_4 *n'est pas compact et si les lignes de temps sont complètes (compactes ou de durée infinie) toutes les feuilles de simultanéité* V_3 *soit isomorphes; si* V_3 *est compacte, alors il n'existe pas de ligne de temps fermée.*

BIBLIOGRAPHIE

1. C. Ehresmann et G. Reeb, Sur les champs d'éléments de contact de dimension *p* complètement intégrables, C.R.A.S., *218* (1944), 995.
2. C. Ehresmann et Shih Weishu, Sur les espaces feuilletés : théorème de stabilité, C.R.A.S., *243* (1956), 344.
3. C. Ehresmann, Espèce de structures locales; Elargissements de catégories, Séminaire Ehresmann (1961) (1ère partie : traduction de Gattungen von Lokalen Strukturen, Jahresbericht Deutsch. Math. Vereinigung, *60* (1957), 49).
4. ——— Introduction à la théorie des structures infinitésimales et des pseudogroupes de Lie, Coll. Int. Géom. Diff. Strasbourg, C.N.R.S. (1953).
5. ——— Catégories inductives et pseudogroupes, Ann. Inst. Fourier, *10* (1960), 307.
6. ——— Structures locales, Annali di Mat. (1954), 133.
7. ——— Catégories topologiques et catégories différentiables, Coll. Geom. Diff. Glob. (Bruxelles, 1958), 137.
8. ——— Sur la théorie des variétés feuilletées, Rend. Mat. e delle sue appl. V, *10*, Fasc. 1–2 (Roma, 1951).
9. ——— Les connexions infinitésimales dans un espace fibré différentiable, Coll. Topologie (Bruxelles, 1950), 29.
10. ——— Les prolongements d'un espace fibré différentiable, C.R.A.S., *240* (1955), 1755.
11. ——— Sur les espaces localement homogènes, l'Enseignement Math., *35*, n. 5–6 (1936), 317.
12. ——— Sur les espaces fibrés associés à une variété différentiable, C.R.A.S., *216* (1943), 628.
13. G. Reeb, Sur certaines propriétés topologiques des variétés feuilletées, ASI 1183 (Paris, 1952).
14. ——— Sur la théorie générale des systèmes dynamiques, Ann. Inst. Fourier, *6* (1955–56), 89.
15. H. Freudenthal, Ueber die Enden topologischer Räume und Gruppen, Math. Zeit. (1931), 692.
16. A. Haefliger, Structures feuilletées et cohomologie à valeur dans un faisceau de groupoïdes, Comm. Math. Helv., *32* (1958), 4.
17. M. H. A. Newman, A theorem on periodic transformations of spaces, Quart. J. Math., *2* (1931), 8.
18. P. A. Smith, Transformations of finite period, III: Newman's theorem, Ann. Math., *42* (1941), 446.
19. D. Montgomery et L. Zippin, Topological transformation groups, Interscience Tracts (New York, 1955).
20. G. De Rham, Com. Math. Helv., *26* (1952), 339.

AN EXTENSION OF THE CONCEPT OF REAL NUMBER*

A. ERDÉLYI, *California Institute of Technology*

1. The classical concept of a real-valued function of a real variable is not sufficiently wide to accommodate the delta function and other "improper" functions. Theories of generalized functions, based on an extension of the concept "function," have been developed and made it possible to fit impulse functions and other improper functions into the framework of mathematical analysis. One such theory, perhaps the most powerful and almost certainly the best known of them all, formed the subject of a course of lectures by L. Schwartz at the Vancouver meeting of the Canadian Mathematical Congress in 1949 **(1)**. I should like to report on a new approach to this problem, an approach that is based on an extension of the concept "real number" (rather than on an extension of the concept "function"). The extended concept of real numbers was proposed by Schmieden and Laugwitz† in 1958 in a paper **(2)** in which the first principles of analysis based on their extended real numbers are also presented; and this analysis is being developed by Laugwitz in a series of papers of which three **(3**; **4**; **5)** have appeared so far. The new analysis has been applied in theoretical physics by Bopp **(6)**, and according to a private communication by Dr. Laugwitz there is reason to hope that the new analysis will be successful in coping with the well-known divergence problems of field theory.

Real numbers form an archimedean ordered field in which a topology is defined by a norm (the absolute value). The generalized real numbers of Schmieden and Laugwitz form a partially ordered

*The preparation of this paper was sponsored partly by the National Science Foundation of the U.S.A. under grant no. G-14002.

†[Added in proof.] Since then A. Robinson developed, from a different point of view, a system of generalized real numbers. A brief summary of Robinson's theory is given in his note in Proc. Nederl. Akad. Wetensch. Ser. A, *64* (1961), 432–40; cf. also W. A. J. Luxemburg, Non-standard analysis, Calif. Inst. Tech. (1962).

ring with a locally convex topology. This increased complexity is balanced by some desirable novel features of which the one crucial from our present point of view is the presence in the new system of infinitely small and infinitely large numbers with the consequent possibility of defining delta functions and other "improper" functions as functions whose variables and values are in the new enlarged system.

The resulting theory is not equivalent to any of the known theories of generalized functions. For instance, the relation $x\delta(x)=0$ which is true for all x in distribution theory does not always hold in the new theory. On the other hand, delta functions cannot be multiplied in distribution theory but they can be multiplied, and their product is a multiple of a delta function, in the Schmieden-Laugwitz theory.

2. Throughout this paper, N will stand for the set of natural numbers (positive integers). Let R be an archimedean ordered field of real numbers x with norm (valuation) $|x|$. The two most important cases are the field of rationals, as in **(2)**, and the field of reals, as in **(3**; **4**; **5)**; but any field between these will do equally well for the algebraic part of the theory; later we shall assume that R is the field of reals.

Cantor's construction of real numbers employs equivalence classes of convergent sequences of rationals, two sequences being regarded as equivalent if their difference is a null sequence. We shall enlarge this construction in two directions: by discarding equivalence and regarding distinct sequences as defining different numbers, and by admitting all, not only the convergent, sequences. The first of these modifications introduces infinitely small numbers and the second, infinitely large numbers.

Accordingly, we define R^*, the system of generalized real numbers, as the set of all (infinite) sequences of elements of R or as the set of all functions on N into R. A typical element of R^* may be written as

$$x = (x_1, x_2, \ldots)$$

where each x_n belongs to R. We shall also write $x_n = \Pi_n x$ and regard Π_n, $n = 1, 2, \ldots$, as a projection operator. $\Omega = (1, 2, 3, \ldots)$ is that element of R^* defined by $\Pi_n\Omega = n$.

For x in R^*, we define $|x|$ as that element of R^* for which $\Pi_n |x| = |\Pi_n x|$. Sum $x + y$, product xy, and the relations $x < y$, $x \leqslant y$ are defined componentwise.

$$\Pi_n(x + y) = \Pi_n x + \Pi_n y,\ \Pi_n xy = (\Pi_n x)(\Pi_n y),\ x < y \text{ if } \Pi_n x < \Pi_n y$$

for all n in N, and similarly for $x \leqslant y$. We also define "weak inequalities" $x <' y$, $x \leqslant' y$:

$x <' y$ if $\Pi_n x < \Pi_n y$ for all but a finite number of n in N, and similarly for $x \leqslant' y$.

If multiplication by elements of R ("scalar multiplication") is defined componentwise, R^* is seen to be a vector space over R as the scalar field. R^* is also a ring which is partially ordered by either of the relations $<$, $\leqslant$ (only by $\leqslant$ if reflexivity of the partial ordering is required). Divisors of zero in R^* are those sequences with at least one zero element. Apart from division by one of these, division is possible.

Those sequences with integer components are called *integers* in R^* and those with rational components, *rationals* in R^*. Every rational in R^* can be represented as the quotient of two integers in R^*.

R can be *embedded* in R^* by assigning to x_0 in R the sequence $(x_0, x_0, x_0, \ldots)$ in R^*. Thus the embedding is described by the following isomorphism ϕ of R into R^*: for x_0 in R, $\phi(x_0)$ in R^* is defined by $\Pi_n \phi(x_0) = x_0$ for all n. We shall usually denote x_0 in R and $\phi(x_0)$ in R^* by the same symbol. This leads to no misunderstanding, since $x_0 x$ is the same element of R^* whether it is interpreted as the scalar product of x_0 in R with x in R^*, or the product $\phi(x_0)x$ in R^*.

Bounded sequences x will be called *finite* elements of R^*. For $x > 0$, $y > 0$ in R^*, we write $x \ll y$ if for each positive α in R, $x <' \alpha y$, that is, if for each $\alpha > 0$ in R there exists an $n_0 = n_0(x, y, \alpha)$ so that $x_n < \alpha y_n$ for $n > n_0$. x in R^* is said to be *infinitely large if* $|x| \gg 1$ and *infinitesimally small* if $|x| \ll 1$. For instance, Ω is infinitely large, and Ω^{-1} is infinitesimally small.

3. We now introduce a *topology* in R^*. Since R^* is a vector space, it will be sufficient to describe a neighbourhood basis for 0. For each $\delta > 0$ in R^* let V_δ be the set of those x in R^* for which $|x| < \delta$.

The collection of the sets V_δ for all $\delta > 0$ in R^* is a neighbourhood basis for 0, thus defining a topology in R^*. The topological space so obtained is a Hausdorff space, so that limits, when they exist, are unique. From now on R^* will denote this topological space.

Suppose that for each $k \in N$, $x_k \in R^*$. The sequence x_k is convergent (R^*) if and only if for each $0 < \delta \in R^*$ there exists a $k_0 = k_0(\delta)$ so that $|x_k - x_l| < \delta$, when $k, l > k_0$. Each projection Π_n is a continuous mapping of R^* onto R. The embedding of R into R^* does not preserve the topology of R. Indeed, if $x_k \in R$ for each $k \in N$, then the sequence x_k is convergent (R^*) if and only if $x_k = x_l$ for all sufficiently large k, l.

For $a, b \in R^*$, we define the *closed interval* as the set of those x in R^* for which $a \leqslant x \leqslant b$, and the *open interval*, by $a < x < b$. In order to obtain non-void sets, $a \leqslant b$ is assumed in the case of the closed interval, and $a < b$ in the case of the open interval. In the latter case we extend the notation, allowing some or all of the components of a to be $-\infty$ and some or all of the components of b to be $+\infty$. $[a, b]$ is a closed set and (a, b) is an open set. We call an interval infinitesimal if $b - a \ll 1$, finite if it is not infinitesimal and if $b - a$ is finite (that is, a bounded sequence), and infinite otherwise.

A new topology in R^* may be characterized by the replacement of $|x| < \delta$ in the definition of V_δ by $|x| <' \delta$. The resulting structure is a topological space but not a Hausdorff space. It becomes a Hausdorff space upon forming equivalence classes, identifying two elements, x and y, of R^* if $x_n = y_n$ for all but a finite set of $n \in N$. This new Hausdorff space, R', is used in (**4**; **5**). At the present stage it is not clear whether R^* or R' is the more suitable space for the applications.

4. From now on we assume that R is the field of real numbers, and R^* is the topological space formed by sequences of real numbers. We wish to discuss functions whose variables and values lie in R^*. The notions of continuity, uniform continuity, and the like for such functions are fully determined by the topology of R^*, and derivatives and integrals can also be defined (**2**, p. 24 f.). In this lecture we shall restrict our attention to a very special class of functions, to be called normal functions, for which the nth component of the value of the function depends only on the nth component of the variable. It turns out that for normal functions derivatives, integrals, and the like can be defined easily in terms of the corresponding concepts for real-valued functions of a real variable.

The *extension* of a *set* D_0 in R is the set D_0^* of all those $x \in R^*$ for which $\Pi_n x \in D_0$ for all n. In this sense R^* itself is the extension of R. The *extension* of a *function* f_0 on $D_0 \subset R$ to R is that function f_0^* defined on $D_0^* \subset R^*$ to R^* by $\Pi_n f_0^* (x) = f_0 (\Pi_n x)$, that is, $f_0^* (x) = (f_0(x_1), f_0(x_2), \ldots)$. It is easy to see that continuity of f_0 on the set $[\Pi_n x : n \in N]$ implies continuity of $f^*{}_0$ at x; and that continuity (uniform continuity) of f_0 on D_0 implies continuity (uniform continuity) of f_0^* on D_0^*.

For instance, the real exponential function e^x can be extended to $e^x = (e^{x_1}, e^{x_2}, \ldots)$ or $\Pi_n e^x = e^{x_n}$. Rational operations are preserved in the process of extension, in particular, $e^x e^y = e^{x+y}$ holds for the extended function. Other elementary functions, and identities involving them, can be extended similarly.

A function f on $D \subset R^*$ to R^* is called a *normal function* if for each $n \in N$ there exists a function f_n on $\Pi_n D \subset R$ to R such that $\Pi_n f(x) = f_n (\Pi_n x)$. Clearly, extensions of real-valued functions of a real variable are normal. There are normal functions which are not extensions: $f(x) = x_1, x_2^2, x_3^3, \ldots)$ defines a normal function which might be denoted by x^Ω, and this function is not the extension of some real-valued function of a real variable.

A normal function f on $D \in R^*$ to R^* is continuous at $x \in D$ (uniformly continuous on D) if and only if each f_n is continuous at x_n (uniformly continuous on $\Pi_n D$). It will be instructive to run through the proof of this statement for continuity of f at x. For this it is necessary and sufficient that for each $0 < \eta \in R^*$ there exist a $0 < \delta \in R^*$ so that $|f(y) - f(x)| < \eta$ whenever $y \in D$ and $|y - x| < \delta$; and this is equivalent to demanding that for each $n \in N$ and each $0 < \eta_n \in R$ there exist a $0 < \delta_n \in R$ so that $|f_n(y_n) - f_n(x_n)| < \eta_n$ whenever $y_n \in \Pi_n D$ and $|y_n - x_n| < \delta_n$; and this in its turn is precisely the necessary and sufficient condition for the continuity of each of the f_n at x_n.

For a normal function $f(x)$ the derivative, $f'(x)$, may be defined by $\Pi_n f'(x) = f_n'(x_n)$, and it turns out that for normal functions this is equivalent to the topological definition according to which $f'(x)$ is the limit, if it exists, of

$$\frac{f(x+h)-f(x)}{h} \quad \text{as} \quad 0<|h|\to 0.$$

In the same spirit, the integral of a normal function over an interval in R^* may be defined by

$$\Pi_n \int_a^b f(x)dx = \int_{a_n}^{b_n} f_n(t)dt.$$

For normal functions the familiar theorems of elementary analysis remain valid and can be established easily componentwise: these include the intermediate value theorem for continuous functions, the continuity of differentiable functions, the integrability of continuous functions, the fundamental theorem of calculus, the mean value theorems, and many others. Likewise, the familiar rules for manipulating derivatives and integrals, including integration by parts and changing the variable of integration, hold.

5. Delta functions can be introduced at varying levels of generality. With all the definitions proposed so far, one obtains a family of delta functions rather than a single function: the various members of the family are close to each other (in a sense to be explained later) but they are not identical. The definition to be given here is more restricted than that proposed in **(3)**.*

Let I_0 be an interval in R, c a point of I_0, and let I be the extension of I_0 to R^*. A function δ on I to R^* is called a *delta function* on I concentrated at (or in the vicinity of) c provided that (i) δ is a normal continuous function on I;

(ii) $$\int_I |\delta(x)|dx \leqslant A \in R, \qquad \int_I \delta(x)dx = 1;$$

and (iii) there is an infinitesimal interval J such that $c \in J \subset I$, $\delta(x) \geqslant 0$ on J, and

$$\int_{I-J} |\delta(x)|dx \ll 1.$$

*__(3)__ contains some inaccuracies which, Dr. Laugwitz informs the writer of the present account, can be removed by using the methods of **(4)** and thus basing the analysis on R' rather than R^*.

We shall show that there are delta functions in the sense of this definition by constructing a master formula which will yield a variety of delta functions. Indeed many, although not all, of the familiar approximations to the delta function are related to special cases of the master formula.

Let γ be a normal continuous function on R^* to R^* and such that $\gamma(x) \geqslant 0$, $\int_{R^*}\gamma(x)dx = 1$. Then for each sufficiently large $c \in R^*$, $\delta(x) = c\gamma(cx)$ is a delta function on R^* concentrated at 0. In particular, if γ is the extension of a function on R to R, δ will be a delta function whenever $c \gg 1$.

To prove this result, first note that there exists $0 < \alpha \in R^*$ so that

$$0 \leqslant \int_{\alpha}^{\infty} \gamma(x)dx \ll 1, \qquad 0 \leqslant \int_{-\infty}^{-\alpha} \gamma(x)dx \ll 1.$$

Now, for $0 < c \in R^*$, $\delta(x) = c\gamma(cx)$ is a normal and continuous function, $\delta(x) \geqslant 0$, and

$$\int_{R^*} \delta(x)dx = \int_{R^*} c\gamma(cx)dx = \int_{R^*} \gamma(y)dy = 1.$$

Let $0 < \beta \ll 1$ and let J be the infinitesimal interval $(-\beta, \beta)$. Now,

$$0 \leqslant \int_{\beta}^{\infty} \delta(x)dx = \int_{\beta c}^{\infty} \gamma(y)dy \ll 1$$

provided $\beta c \geqslant \alpha$, and similarly

$$\int_{-\infty}^{-\beta} \delta(x)dx \ll 1$$

in this case. There exists a β satisfying both $0 < \beta \ll 1$ and $\beta c \geqslant \alpha$ provided $c \gg \alpha$, and for c sufficiently large in this sense, $c\gamma(cx)$ is a delta function. If γ is the extension of a real function, $c \gg 1$, $\alpha = c^{\frac{1}{2}}$, $\beta = c^{-\frac{1}{2}}$ satisfies all conditions.

6. The most important property of the delta function is the *reproducing* or *sifting property* which asserts that under certain conditions on f,

$$\int_I f(x)\delta(x)dx$$

reproduces $f(c)$, provided that δ is a delta function on I concentrated at c. Although this property can be established under more general circumstances, we shall content ourselves with showing that for a delta function δ on I concentrated at c,

$$\left| \int_I f(x)\delta(x)dx - f(c) \right| \ll 1$$

provided that f is the normal extension of a real-valued bounded continuous function on I_0.

We wish to show that

$$\left| \int_I [f(x) - f(c)]\delta(x)dx \right| \ll 1,$$

given that $|f(x)| \leqslant M \in R$. By property (iii) of the delta function

$$\int_{I-J} |f(x) - f(c)|\, |\delta(x)|\, dx \leqslant 2M \int_{I-J} |\delta(x)|\, dx \ll 1.$$

It remains to show that also

$$\left| \int_J [f(x) - f(c)]\delta(x)dx \right| \ll 1.$$

Now, $f(x) - f(c)$ and $\delta(x)$ are normal continuous functions, and $\delta(x) \geqslant 0$, so that by the mean value theorem

$$\int_J [f(x) - f(c)]\delta(x)dx = [f(\tilde{c}) - f(c)] \int_J \delta(x)dx,$$

where $\tilde{c}$ is some point of J. We then have

$$\left| \int_J \delta(x)dx \right| \leqslant \int_I |\delta(x)|\, dx \leqslant A \in R$$

and, J being an infinitesimal interval, $|\tilde{c} - c| \ll 1$ and hence $|f(\tilde{c}) - f(c)| \ll 1$. This completes the proof.

Let a and b be the end-points of I, and let a_1 and b_1 be the end-points of J. The function

$$h(x) = \int_a^x \delta(t)dt$$

corresponds to Heaviside's unit function. h is a bounded and continuous normal function on I;

$$|h(x)| \leqslant \int_{I-J} |\delta(x)|\, dx \ll 1 \quad \text{if} \quad x \leqslant a_1;$$

h is monotonic increasing on J (in the sense of the partial ordering of R^*); and

$$|h(x) - 1| \leqslant \int_{I-J} |\delta(x)|\, dx \ll 1 \quad \text{if} \quad x \geqslant b_1.$$

7. We have seen that for fixed I and c there is an infinity of delta functions on I concentrated at c and we may ask if all these delta functions are close to each other in some sense.

Let δ_1 and δ_2 be the two delta functions on I concentrated at c, let A_1 and A_2 be the corresponding bounds occurring in condition (ii) of § 5, let J_1 and J_2 be the infinitesimal intervals of concentration, let J be an infinitesimal interval containing both J_1 and J_2 (we may take $\Pi_n J = \Pi_n J_1 \cup \Pi_n J_2$), and let h_1 and h_2 be the corresponding "unit functions" defined by

$$h_i(x) = \int_a^x \delta_i(t)dt, \qquad i = 1, 2.$$

It follows from the estimates given at the end of the preceding section that $|h_1(x) - h_2(x)| \ll 1$ if $x \in I - J$, but on J, $h_1(x) - h_2(x)$ need not be small, indeed $|h_1(x) - h_2(x)| \leqslant A_1 + A_2 = A$, say is about the best estimate holding generally on J.

Let us now consider

$$k_i(x) = \int_c^x h_i(t)dt.$$

We shall show that $|k_1(x) - k_2(x)| \ll 1$ for all finite x, in particular, for all x in I if x is a finite interval. If $x \in J$, then

$$|k_1(x) - k_2(x)| = \left| \int_c^x [h_1(t) - h_2(t)]dt \right| \leqslant A\, |x - c| \ll 1$$

since $A \in R$ and $|x - c| \ll 1$ when $x \in J$. Let a_1, b_1 be the end-points of J and assume that x is finite and $x \leqslant a_1$, say. (A similar proof holds for $x \geqslant b_1$.) Then

$$|h_1(x) - h_2(x)| \leqslant \int_{I-J} |\delta_1(x) - \delta_2(x)|\, dx \ll 1, x < a_1$$

and

$$|k_1(x) - k_2(x)| \leqslant \left| \int_x^{a_1} [h_1(t) - h_2(t)]dt \right| + \left| \int_{a_1}^{c} [h_1(t) - h_2(t)]dt \right|$$

$$\leqslant |x - a_1| \int_{I-J} |\delta_1(t) - \delta_2(t)|\, dt + A\, |c - a_1| \ll 1$$

since A and $|x - a_1|$ are finite while $\int_{I-J}$ and $|c - a_1|$ are infinitesimally small.

Thus we see that, although the delta function is not determined by I and c, the repeated integral

$$k(x) = \int_c^x \left(\int_a^t \delta(s)ds \right) dt.$$

is determined up to a function which is infinitesimally small for finite x, indeed uniformly in x on each finite subinterval of I.

8. The circumstances described in the preceding section suggest an approach, within the Schmieden-Laugwitz theory, to distributions.

In the class of all normal functions on I let us regard two functions, f and g, as equivalent provided that for each closed finite subinterval J of I there exist a non-negative integer m and two m times differentiable normal functions F and G such that $f = F^{(m)}$, $g = G^{(m)}$, and $|F(x) - G(x)| \leqslant M \ll 1$ on J. Clearly this notion of equivalence is reflexive and symmetric, and it is easily shown to be transitive. It divides all normal functions on I into equivalence classes.

Laugwitz **(4)** now defines distributions on I as equivalence classes of normal functions on I, and relates his definition to that given by Mikusinski and Sikorski **(7)**.

In this sense all delta functions on a fixed interval I and concentrated at a fixed point c of I are equivalent. They determine a unique distribution on I, called the *delta distribution*. Thus, the delta function appears in the Schmieden–Laugwitz theory either as a distribution which is unique but does not possess functional values, or else as a function which is highly non-unique but has the advantage of possessing functional values. This flexibility has proved useful in the applications (**6**).

We conclude with a direct characterization of the delta distribution which is of interest since it corresponds very closely to the intuitive notion of the delta function as the derivative of Heaviside's unit function. Let $H(x)$ be the normal extension of Heaviside's unit function, that is, let H be the normal function on R^* defined uniquely by

$$H(x) = \begin{matrix} 0 \text{ if } x < 0 \\ \\ 1 \text{ if } x = 0 \text{ or } x > 0. \end{matrix}$$

Then the delta distribution for the interval I and concentrated at c is the equivalence class on I containing all those integrable normal functions f on I for which $\int_a^x f(t)dt$ is equivalent, in the sense explained above, to $H(x - c)$. This approach can be used to give a definition of delta functions that is less restrictive than the one given in § 5, and also in order to prove the sifting property for a more general class of functions than that envisaged in § 6.

REFERENCES

1. I. Halperin, Introduction to the theory of distributions: based on lectures by Laurent Schwartz (Toronto, 1952).
2. C. Schmieden and D. Laugwitz, Eine Erweiterung der Infinitesimalrechnung, Math. Zeitsch., *69* (1958), 1–39.
3. Detlef Laugwitz, Eine Einführung der δ-Funktionen, Sitzungsber. Bayr. Akad. Wiss. (1959), 41–59.
4. ——— Anwendungen unendlich kleiner Zahlen I: Zur Theorie der Distributionen, J. reine angew. Math., *207* (1961), 53–60.
5. ——— Anwendungen unendlich kleiner Zahlen II: Ein Zugang zur Operatorenrechnung von Mikusinski, J. reine angew. Math., *208* (1961), 22–34.
6. Fritz Bopp, Lorentzinvariante Wellengleichungen für Mehrbahnsysteme, Sitzungsber. Bayr., Akad. Wiss. (1958), 167–225.
7. J. Mikusinski and R. Sikorski, The elementary theory of distributions, I. Rozprawy matematyczne, *12* (1957).

MATHEMATICS AND LOGIC

LEON HENKIN, *University of California, Berkeley*

It was 24 years ago this month that I entered Columbia College as a freshman and discovered the subject of logic. I can recall well the particular circumstance which led to this discovery. One day I was browsing in the library and came across a little volume by Bertrand Russell entitled *Mysticism and logic.* At that time, barely sixteen, I fancied myself something of a mystic. Like many young people of that age I was filled with new emotions strongly felt. It was natural that my reflective attention should be largely occupied with these, and that this preoccupation should give a colour and poignancy to experience which found sympathetic reflection in the writings of men of mystical bent.

Having heard that Russell was a logician I inferred from the title of his work that his purpose was to contrast mysticism with logic in order to exalt the latter at the expense of the former, and I determined to read the essay in order to refute it. But I discovered something quite different from what I had imagined. Indeed, contrasting aspects of mysticism and logic were delineated by Russell, but his thesis was that each had a proper and important place to play in the totality of human experience, and his interest was to define these and to exhibit their interdependence, rather than to select one as superior to the other. I was disarmed, I was delighted with Russell's lucent and persuasive style, I began avidly to read his other works, and was soon caught up with logical concepts which have continued to occupy at least a portion of my attention ever since.

Bertrand Russell has popularized a good many ideas, abstract as well as concrete. Probably many of you have been afforded an introduction to mathematical logic through his writings, and perhaps some have even been led to the point of peeping into the formidable *Principia Mathematica* which he wrote with Alfred Whitehead about 1910. You will recall, then, the astonishing contention with which he shocked the mathematical world of that time, namely, that all of mathematics was nothing but logic. Mathematicians were generally puzzled by this radical thesis. Though few understood what Russell had in mind, nevertheless, they vehemently opposed the idea.

This is readily understandable when one recalls that a companion thesis of Russell's was that logic is purely tautological, and has really no content whatever. Mathematicians, being adept at putting 2 and 2 together, quickly inferred that Russell meant to say that all mathematical propositions are completely devoid of content, and from this it was a simple matter to pass to the supposition that he held all mathematics to be entirely without value. *Aux armes, citoyens du monde mathématique!*

Half a century has elapsed since this gross misinterpretation of Russell's provocative enunciation. These fifty years have been host to a great acceleration and broadening of logical research. And so it seems to me appropriate, on this occasion, to seek a reassessment of Russell's thesis in the light of subsequent development.

In order to explain how Russell came to hold the view that all of mathematics is nothing but logic, it is necessary to go back and discuss two important complexes of ideas which had been developed in the decades before Russell came into the field. The first of these was a systematic reduction of all the concepts of mathematics to a small number of them. This process of reduction had indeed been going on for a very long time. As far back as the days of Descartes, for example, we can see at least an imperfect reduction of geometric notions to algebraic ones. Subsequently, with the development of set theory initiated by Cantor, the reduction of the real number system to that of the natural numbers marked another great step in this process. But perhaps the most daring culmination of these efforts was the attempt by a German mathematician, Gottlob Frege, to analyse the notion of natural number still further and reduce it to a concept which he considered to be of a purely logical nature.

Frege's work was almost entirely unnoticed in his own time (the last three decades of the nineteenth century), but when Bertrand Russell came upon Frege's work he realized its great significance and gave these ideas wide currency through his own brilliant style of exposition. The ultimate elements into which the notion of natural number was analysed by Frege and by Russell consisted of entities which they called "propositional functions." Now propositions had long been recognized as constituting one of the most basic portions of the domain of investigation of logicians and, because propositional functions were very closely related to propositions themselves, it was natural to consider these too to be a proper part of the subject of logic. It is in this sense that Frege seemed able by a series of definitions to arrive at the notion of number, as well as at the other notions under study in various parts of mathematics, starting from purely logical notions.

The second important line of development which preceded Russell, and upon which he drew for his ideas, was the systematic study by mathematical means of the laws of logic which entered into mathematical proofs. This development was initiated by George Boole, working in England in the middle of the nineteenth century. He discovered that certain of the well-known laws of logic could be formulated with the aid of algebraic symbols such as the plus sign, the multiplication sign, the equality sign, and variables. And he pointed out that through the use of this algebraic type of notation one could expect a great saving in the effort needed to collate and apply basic laws of logic. Later his work was extended and deepened by the German mathematician, Schröder. And Russell himself, working within this tradition, found it a convenient basis for a systematic development of all mathematics from logic. By combining the symbolic formulation of logical laws with the reduction of mathematical concepts to a logical core, he was able to conceive of a unified development such as was attempted in the *Principia Mathematica*.

What was the *Principia* like? The work is still not completed, and because Russell has most recently seemed to occupy himself with the effects of certain physical research it may, perhaps never be completed. Nevertheless, one can see clearly the intended scope of the work. Surprisingly, it reminds one of the present massive undertaking by the Bourbaki group in France. For even though the *Principia* and Bourbaki are dissimilar in many ways, each attempts to present an encyclopaedic account of contemporary mathematical research unified by a coherent point of view.

In the *Principia*, starting from certain axioms expressed in symbolic form which were intended to express certain laws of logic (axioms involving only what Russell conceived to be logical notions), the work systematically proceeds to derive the other laws of logic, to introduce by definition such mathematical notions as the concept of number, of geometric space, and so on, and finally to develop the main theorems concerning these concepts as part of a uniform and systemic development.

In retrospect, the contemporary logician is struck by the willingness of Russell and Whitehead to rest their case on what, for a mathematician, must be considered flimsy evidence. The world of empirical science, of course, expects to achieve conviction on the basis of empirical evidence, but the quintessence of the mathematician's approach, especially of the mathematical logician's, is the demand always for *proof* before a thesis is accepted. Yet, whereas Russell was interested in establishing that in a certain sense all of mathematics could be obtained from his logical axioms and concepts, he never really set out to give a proof of this fact! All he did was to gather the basic ideas that had been developed in a non-formal and unsystematic way by mathematicians before him, and to say, in effect, "You see that I have been able to introduce all this loosely formulated work within the precise framework of my formal system. And it's pretty clear, isn't it, that I have all the tools available to formalize such further work as mathematicians are likely to do?"

In this respect one is reminded of the approach of that first great axiomatizer and geometer, Euclid. Euclid, too, conceived that all propositions of geometry, that is, all the true statements about triangles, circles, and those other figures in which he was interested, could be developed from the simple list of concepts and axioms he gave. But in his case, too, there was never any attempt to prove this fact other than by the empirical process of deriving a large number of geometric propositions from the axioms and then appealing to the good will of the audience, so to speak. "Well," we may imagine him saying, "look how much I have been able to deduce from my axioms. Aren't you pretty well convinced that *all* geometric facts follow from them?" But of course there were mathematicians and logicians who were *not* convinced. And so the demand for proof was raised.

Actually, the proper formulation of the problem whether a system of axioms is adequate to establish all of the true statements in some domain of investigation requires a mathematically precise formulation of the notion of "true sentence," and it was not until 1935 that Tarski, in a great pioneering work, made fully evident the form in which semantical notions must be analysed for mathematical languages. But much earlier than that, as early as 1919, we find the first proof of what we call in logic *completeness*. The mathematician Emil Post (in his doctoral disseration published in that year), limiting his attention to a very small fragment of the system created by Whitehead and Russell, was able to show that for any sentence in that fragment which was "true under the intended interpretation of the symbols," one could indeed get a proof by means of the axioms and rules of inference which had been stated for the system. Subsequently, further efforts were made to extend the type of completeness proof which Post initiated, and it was hoped that ultimately the entire system of the *Principia* could be brought within the scope of proofs of this kind. In 1930, Kurt Gödel contributed greatly to this development and to this hope when he succeeded in proving the completeness of a deductive system based upon a much larger portion of mathematical language than had been treated by Post. And indeed, Gödel's proof of the completeness of the so-called first order predicate logic is among the basic and useful theorems we have in the whole subject of mathematical logic.

But the very next year, in 1931, the hope of further extension of this kind of completeness proof was definitely dashed by Gödel himself in what is certainly the deepest and most famous of all works in mathematical logic. Gödel was able to demonstrate that the system of *Principia Mathematica*, taken as a whole, was *incomplete*. That is, he showed explicitly how to construct a certain sentence about natural numbers, which mathematicians could recognize as being true under the intended interpretation of the symbolism, but which could not be proved from the axioms by the rules of inference which were part of that system.

Now, of course, if Gödel had done nothing more than this, one might simply conclude that Russell and Whitehead had been somewhat careless in formulating their axioms, that they had left out this true but unprovable sentence from among the axioms, and one might hope that by adding it as a new axiom a stronger system which was complete would be achieved. But Gödel's proof shows that this stronger system, too, would contain a sentence which is true but not provable; that indeed if this system were further strengthened, by the addition of this new true but unprovable sentence as an axiom, the resulting system would again be incomplete. And indeed, if a whole infinite sequence of sentences were to be obtained by successive applications of Gödel's method, and added simultaneously to the original axioms of *Principia*, the same process could still be applied to find *another* true sentence still unprovable.

Actually, Gödel described a very wide class of formal deductive systems to which his method applies. And most students of the subject have been convinced that any formal system of axioms and rules of inference which it would be reasonable to consider as a basis for a development of mathematics would fall in this class, and hence would suffer a form of incompleteness. From this viewpoint it appears as if one of the basic elements on which Russell rested his thesis that all mathematics could be reduced to logic must be withdrawn and reconsidered.

I have been talking about completeness, which has to do with the adequacy of a formal system of axioms and rules of inference for proving true sentences. But I must mention a second aim of the Russell–Whitehead *Principia* which also fared ill in the subsequent development of mathematical logic. Russell and Whitehead were very much concerned with the question of *consistency*. While they hoped to have a complete system, one containing proofs for all correct statements, they were also concerned that their system should *not* contain proofs of *incorrect* results. In particular, in a consistent system such as they sought, it would not be possible to prove both a sentence and its negation.

To understand their concern with the question of consistency it is necessary to recall the rude awakening of mathematicians in 1897 with Cantor's theory of transfinite numbers. For centuries before the time of Cantor mathematicians simply assumed that any one who was properly educated in their subject could distinguish a correct proof from an incorrect one. Those who had trouble in making this distinction were simply "weeded out" in the course of their training and were turned from mathematics to lesser fields of study. And no one took up seriously the question of setting forth, in explicit and mathematical terms, exactly what was meant by a correct proof.

Now when Cantor began his development of set theory he concerned himself with both cardinal and ordinal numbers of transfinite type. (These numbers can be used for infinite sets in very much the same way as we use ordinary numbers for counting and ordering finite sets.) Many of the properties of transfinite numbers are identical to those of ordinary numbers, and in particular Cantor showed that given any ordinal number b we can obtain a larger number, $b + 1$. However, in 1897 an Italian mathematician, C. Burali-Forti, demonstrated that there must be a *largest* ordinal number, by considering the set of all ordinal numbers in their natural order. Mathematicians were unable to find any point, either in the argument of Cantor or of Burali-Forti, which seemed intuitively to rest on incorrect reasoning. Gradually it was realized that mathematicians had a genuine paradox on their hands, and that they would have to grapple at last with the question as to just what was meant by a correct proof. Later, Russell himself produced an even simpler paradox in the intuitive theory of sets, based upon the set of all those sets which are not elements of themselves.

This background sketch will make clear why Russell and Whitehead were much concerned that no paradox should be demonstrable in their own system. And yet they themselves never attempted a *proof* that their system was consistent! Their only evidence was that a large number of theorems had been obtained within their system without encountering paradox, and that all attempts to reproduce within the system of *Principia Mathematica* the Burali-Forti paradox, and such other paradoxes as were known, had failed.

As with the question of completeness, mathematicians were not satisfied with an answer in this form, and there arose a demand that an actual *proof* of the consistency be given for the system of *Principia* (and for other systems then considered). The great and illustrious name of Hilbert was associated with these efforts to achieve consistency proofs for various portions of mathematics, and under his stimulus and direction important advances were made towards this goal, both by Hilbert and by his students. But as with the efforts to prove completeness, Hilbert's programme foundered upon the brilliant ideas of Kurt Gödel.

Indeed, in that same 1931 paper to which we have previously referred, Gödel was able to show that the questions of consistency and completeness were very closely linked to one another. He was able to show that *if* a system such as the *Principia* were truly consistent, then it would not be possible to produce a sound proof of this fact! Now this result itself sounds paradoxical. Nevertheless, when expressed with the technical apparatus which Gödel developed it is a precisely established and clearly meaningful mathematical result which has persuaded most, though admittedly not all, logicians that Hilbert's search for a consistency proof must remain unfulfilled.

I should like finally to mention a third respect in which the original aim of mathematical logicians were frustrated. The questions of consistency and completeness, mentioned above, clearly concerned the authors of *Principia Mathematica*; but the question of *decision procedures* seems not to have been treated to any serious extent by Russell and Whitehead. Nevertheless, this is an area of study which interested logicians as far back as the time of Leibniz. Indeed Leibniz himself had a dream that it might be possible to devise a systematic procedure for answering questions—not only mathematical questions but even questions of empirical science as well. Such a procedure was to replace the need for inspiration with the automatic carrying out of routine procedure. Had Leibniz been conversant with today's high speed computing machines, he might have formulated his idea by asserting the possibility that one could write a programme for a machine of such breadth and inclusiveness that any scientific question whatever could be placed on tape for which, after the machine had been set to work on it for some finite length of time, a definitive reply would be forthcoming.

Leibniz' idea lay dormant for a long time; but it was natural to revive it in connection with the formal deductive systems which were developed by mathematical logicians in the early part of this century. Because these logicians had been interested in formulating mathematical ideas within a symbolic calculus, and then manipulating the symbols according to predetermined rules in order to obtain further information about these mathematical concepts, it seemed natural to raise the question whether one could not devise purely automatic rules of computation which would enable one to reach a decision as to the truth or falsity of any given sentence of the calculus. And though the area of empirical science was pretty well excluded from the consideration of twentieth-century logicians seeking such decision procedures, it was perhaps not beyond the hopes of some that a system as inclusive as that of the *Principia* could some day be brought within the scope of such a procedure.

Efforts to find decision procedures for various fragments of the *Principia* were vigorous and many. The doctoral dissertation of Post, for example, contained some efforts in this direction, and further work was produced during the succeeding 15 years by logicians of many countries. Then in 1936 Alonzo Church, making use of the newly developed notion of recursive function, was able to demonstrate that for a certain fragment of mathematical language, in fact for that very first-order predicate logic which Gödel had showed to be complete in 1930, no decision procedure was possible. And so with decision procedures, as with proofs of completeness and consistency, efforts to establish a close rapport between logic and mathematics came to an unhappy end.

Probably most mathematicians have heard at least something of the development which I have sketched here. But the logical education of most mathematicians seems to have been terminated at about that point. The impression is fairly widespread that the discoveries of Gödel and Church effectively ground to a halt the ambitious programme of mathematical logicians, and that since then further work in logic has been confined to a sort of helpless faltering by people, unwilling to accept the cruel facts of life, who are still seeking to buttress the advancing frontiers of mathematical research by finding a non-existent consistency proof.

This image is very far indeed from reality. For in 1936, just when many suppose the demise of mathematical logic to have been completed, an international scholarly society known as the Association for Symbolic Logic was founded and began publication of the *Journal of Symbolic Logic.* In the ensuing 25 years the *Journal* has greatly expanded to accommodate a growing volume of research. And at present there are four journals devoted exclusively to publishing material dealing with mathematical logic, while many articles on logic appear in a variety of mathematical journals of a less specialized nature.

It may be interesting now to review briefly some of the developments in mathematical logic since 1936. I have found it convenient for this exposition to divide research in mathematical logic into seven principal areas.

First I shall mention the area dealing with the foundations of the theory of sets. To explain the connection of this field with logic it should be mentioned that those objects which Russell and Whitehead had called *propositional functions* are, in fact, largely indistinguishable from what are now called *sets* and *relations* by mathematicians. From a philosophical point of view there is perhaps still room for distinguishing these concepts from one another. But, because the treatment of propositional functions in *Principia Mathematica* is extensional (so that two functions which apply to exactly the same objects are never distinguished), for mathematical purposes this system is identical to one which treats sets and relations.

Among systems of set theory which have been put forth by logicians as a basis for the development of mathematics, the principal ones are the theory of types used by Whitehead and Russell themselves, subsequently simplified by Chwistek and Ramsey, and an alternative line of development initiated by Zermelo to which important contributions were subsequently made by Fraenkel and Skolem. Still another system, having certain characteristics in common with each of these two principal forms, was advanced and has been studied by Quine and, to some extent, by Rosser. Of these systems the Zermelo type has probably received most attention, along with an important variant form suggested and developed by von Neumann, Bernays, and Gödel.

Among the significant efforts expended on these systems were those directed towards establishing the status of propositions such as the Axiom of Choice and the Continuum Hypothesis of Cantor. Here the names of Gödel and Mostowski are especially prominent. Gödel showed that a strong form of the Axiom of Choice and the Generalized Continuum Hypothesis are simultaneously consistent with the more elementary axioms of set theory—under the assumption that the latter are consistent by themselves. Mostowski showed that the Axiom of Choice is independent of the more elementary axioms of set theory—provided that a form of these elementary axioms is selected which does not exclude the existence of non-denumerably many "Urelemente" (objects which are not sets). The independence of the Axiom of Choice from systems of axioms such as that used in Gödel's consistency proof, and the independence of the Continuum Hypothesis in any known system of set theory, remain open questions.[1]

More recently the direction of research in the area of foundations of set theory seems to have shifted from that of formulating specific axiom systems and deriving theorems within them, to the consideration of the totality of different realizations of such axiom systems. It is perhaps Shepherdson who should be given credit for the decisive step in this shift of emphasis, although his work clearly owes much to Gödel's. Subsequent work by Tarski, Vaught, and Montague has carried this development much further.

An important tool in their work is Tarski's concept of the *rank* of a set, which may be defined inductively as the least ordinal number exceeding the rank of all elements of the set. This notion may be used to classify models of set theory according to the least ordinal number which is not the rank of some set of the model. Recently there have been some very interesting contributions by Azriel Levy to these studies. His efforts have been directed towards successively strengthening the axioms of set theory so as to penetrate increasingly far into the realm of the transfinite.

A second area that I would delineate in contemporary logical research is that dealing with the decision problem. While it is true that the work of Church made clear that there could be no *universal* decision procedure for mathematics, there has remained a strong interest in finding decision procedures for more modest portions of mathematical theory. Of special interest here is Tarski's decision method for elementary algebra and geometry, and an important extension of it which was made by Abraham Robinson. Wanda Szmielew has also given an important decision procedure, namely one for the so-called "elementary theory" of Abelian groups. By

[1](Added June, 1963.) Within the last month these questions have been resolved by Paul Cohen, who demonstrated for a system such as Gödel's that the Axiom of Choice is not provable from the other axioms, and that when this axiom is added the Continuum Hypothesis is still not provable. A brief account of these findings is to be reported in the *Proceedings of the National Academy of Sciences*.

contrast, the elementary theory of *all* groups was shown by Tarski to admit *no* decision procedure. In fact, Szmielew and Tarski considered exactly the same set of sentences, roughly all of those sentences which can be built up by the use of the group operation symbol, and variables ranging over the group elements, with the aid of the equality sign, as well as the usual logical connectives and quantifiers. If we ask whether any given sentence of this kind is true for all Abelian groups, it is possible to answer the question in an automatic way by using the method of Szmielew. But if we are interested in which of these sentences are true for all groups, then Tarski's proof shows that it is impossible to devise a machine method to separate the true from the false ones.

A result closely related to Tarski's is that of Novikov and Boone concerning the non-existence of a decision method which would enable one to solve the word problem for the theory of groups, a problem for which a solution had long been sought by algebraists. In fact it is a simple matter to show that the Novikov-Boone result is equivalent to the non-existence of a decision method for a certain subset of the sentences making up the elementary theory of groups, namely all those sentences having a special, very simple, form. Hence this result is stronger than Tarski's.

Now the key concept whose development was needed before negative solutions to decision problems could be achieved was the concept of a recursive function. Intuitively speaking this is simply a function from natural numbers to natural numbers which has the property that there is an automatic method for computing its value for any given argument. A satisfactory and explicit mathematical definition of this class of functions was first formulated by Herbrand and Gödel. But it remained for Kleene to develop the concept to such an extent that it now underlies a very large and important part of logical research.

Much of the work with recursive functions has been along the line of classifying sets and functions, a classification similar to that involving projective and analytic sets in descriptive set theory. Kleene himself, his students Addison and Spector, and other logicians including Mostowski, Shoenfield, and Kreisel have contributed largely to this development. Also to be mentioned are the applications which initially Kleene, and subsequently others, have attempted to make of the concept of recursive function, by way of explicating the notion of "constructive" mathematical processes. In this connection several attempts have been made to link the notion of recursive function with the mathematical viewpoint known as intuitionism, a radical reinterpretation of mathematical language which was advanced by Brouwer and developed by Heyting.

A fourth area of logical research deals with material which has recently been described as algebraic logic. This is a development which can be traced back to the very early work of Boole and Schröder. However, interest in the subject has shifted away from the formulation and derivation of algebraic equations which express laws of logic to the consideration of abstract structures which are defined by means of such equations. Thus the theory of Boolean algebras, relation algebras, cylindric and polyadic algebras have all successively received attention; Stone, Tarski, and Halmos are closely associated with the central development here. The algebraic structures studied in this domain may be associated in a natural way with mathematical theories, and this association permits the use of very strong algebraic methods in the metamathematical analysis of these theories.

A fifth area of modern logical research concerns the so-called theory of models. Here effort is directed towards correlating mathematical properties possessed by a class of structures defined by means of given mathematical sentences, with the structural properties of those sentences themselves. A very early example is Garrett Birkhoff's result that for a class of structures to be definable by means of a set of equational identities it is necessary and sufficient that it be closed under formation of substructures, direct products, and homomorphic images. Characterizations of a similar nature were given for classes definable by universal elementary sentences (Tarski), and by any elementary sentences (Keisler).

A related type of result is Lyndon's theorem that any elementary sentence whose truth is preserved under passage from a model of the sentence to a homomorphic image of that model, must be equivalent to a sentence which does not contain negation signs. In a different direction, Beth has shown that if a given set symbol or relation symbol is not definable in terms of the other symbols of an elementary axiom system, then there must exist two distinct models of these axioms which are alike in all respects except for the interpretation of the given symbol. (This proves the completeness of Padoa's method of demonstrating non-definability.) A logical interpolation theorem of Craig's provides a close link for the results of Lyndon and Beth.

A sixth area which can be discerned in recent work on logic concerns the theory of proof. This is perhaps the oldest and most basic portion of logic, seeking to find systematic rules of proof, or deduction, by means of which the consequences of any propositions could be identified. In recent work, however, logicians have begun to depart in radical ways from the type of systems for which rules of proof were originally sought. For example, several attempts have been made to provide rules of proof for languages containing infinitely long formulae, such as sentences with infinitely many disjunctions, conjunctions, and quantified variables. Tarski, Scott, Karp, Hanf, and others have participated in such efforts. Curiously enough, though this direction of research seems at first very far removed from ordinary mathematics, one of the important results was used by Tarski to solve a problem concerning the existence of measures on certain very large spaces which had remained unsolved for many years.

The last area of logical research is a kind of converse study to what we have called algebraic logic. In algebraic logic we are interested in applying methods of algebra to system of logic. But there are also studies in which results and methods of logic are used to establish theorems of modern algebra. The first to have made such applications seems to have been the Russian mathematician Malcev, who in 1941 indicated how the completeness theorem for first-order logic could be used to obtain a theorem on groups. Subsequently the same technique was used by Tarski to construct various non-archimedean ordered fields. Perhaps the best known name in this area is that of Abraham Robinson, formerly associated with the University of Toronto. Among his contributions was the application of logical methods and results to improve a solution, given in 1926 by Artin, to Hilbert's seventeenth problem (of the famous list presented in his address to the 1900 International Congress of Mathematicians). Robinson showed that when a real polynomial which takes only non-negative values is represented as a sum of squares of rational functions, the number of terms needed for the representation depends only on the degree and number of variables of the given polynomial, but is independent of the particular coefficients.

I hope that this brief sketch of some of the areas of contemporary logical research will give an idea of the ways in which logicians have reacted to the theorems of Gödel and Church which, in the period from 1931 to 1936, dealt so harshly with earlier hopes. Speaking generally, one could describe their reaction as compounded of an acceptance of the impossibility of realizing the original hopes for mathematical logic, a relativization of the original programme of seeking completeness and consistency proofs and decision methods, an incorporation of the new methods and constructs which appeared in the impossibility proofs, and the development of quite new interests suggested by generalization of early results.

With this background, let us now return to Russell's thesis that all of mathematics can be reduced to logic. I would say that if logic is understood clearly to contain the theory of sets (and this seems to be a fair account of what Russell had in mind), then most mathematicians would accept without question the fact that the basic concepts of all mathematics can be expressed in terms of logic. They would agree, too, that the theorems of all branches of mathematics can be derived from principles of set theory, although they would recognize that no fixed system of axioms for set theory is adequate to comprehend all of those principles which would be regarded as "mathematically correct."

Perhaps of greater significance is the consensus of mathematicians that there is much more to their field than is indicated by such a reduction of mathematics to logic and set theory. The fact that certain concepts are selected for investigation, from among all logically possible notions definable in set theory, is of the essence. A true understanding of mathematics must involve an explanation of which set-theoretical notions have "mathematical content," a question manifestly not reducible to a problem of logic—however, broadly conceived.

Rather than logic being all of mathematics, it seems to be but one branch. But it is a vigorous and growing branch, and there are reasons to hope that it may in time provide an element of unity to oppose the fragmentation which seems to beset contemporary mathematics—and indeed every branch of scholarship.

BIBLIOGRAPHICAL NOTE

Some of the early work of Boole, Frege, and Cantor has been made available by relatively recent translations into English. For these see (**1**; **3**; and **2**). The work of Russell and Whitehead (**10**), while largely devoted to a formidable formalism, contains many very readable sections of an introductory or summary character. An account of the fundamental theorems of Gödel and Church may be found in (**8**).

In describing logical research since 1936 our aim has not been to give a complete history, but only to indicate the range of activity by mentioning some of the most actively cultivated areas. Accordingly, we content ourselves here by referring only to a small sample of the literature, selecting (where possible) works which are largely self contained. For an example of an axiomatic theory of sets and an important contribution toward its meta-theory, see (**4**). For an account of the decision problem consult (**11**) for positive solutions and (**13**) for negative. The theory of recursive functions and their uses is well described in (**8**). Some historical remarks on algebraic logic, as well as detailed results, can be found in (**6**). A very recent and important contribution to the theory of models is (**7**), where reference to earlier works may be found. For an account of recent work on infinitely long formulae see (**5**); an application of such work to a problem on the existence of certain measures appears in (**12**). Applications of logic to algebra are described in (**9**).

REFERENCES

1. G. Boole, An investigation of the laws of thought (New York, 1951). (Reprinted by Dover Publications.)

2. G. Cantor, Contributions to the founding of the theory of transfinite numbers, trans. Ph. Jourdain (Chicago and London, 1915).

3. G. Frege, The foundations of arithmetic, trans. J. L. Austin (Oxford and New York, 1950).

4. K. Gödel, The consistency of the axiom of choice, Ann. Math. Study No. 3 (Princeton, 1940).

5. L. Henkin, Some remarks on infinitely long formulas *in* Infinitistic methods (New York, 1961).

6. L. Henkin and A. Tarski, Cylindric algebras *in* Lattice Theory, Proc. Symp. Pure Mathematics, vol. II (Providence, 1961).

7. H. J. Keisler, Ultraproducts and elementary classes *in* Indagationes Mathematicae, *23* (1961), 477–495.

8. S. C. Kleene, Introduction to Metamathematics (New York, 1952).
9. A. Robinson, Complete theories, "Studies in Logic" series (Amsterdam, 1956).
10. B. Russell and A. N. Whitehead, Principia Mathematica, *1*, *2*, *3* (Cambridge, 1910, 1912, 1913).
11. A. Tarski, A decision method for elementary algebra and geometry (Berkeley and Los Angeles, 1951).
12. ——— Some problems and results relevant to the foundations of set theory, *in* Proceedings of the International Congress for Logic, Methodology, and Philosophy of Science, Stanford, 1960 (Stanford, 1962).
13. A. Tarski, A. Mostowski, and R. Robinson, Undecidable theories, "Studies in Logic" series (Amsterdam, 1953).

THE APPLICATION OF MATHEMATICS TO BIOLOGY AND MEDICINE

IAN N. SNEDDON, *University of Glasgow*

1. In giving a lecture to a congress of this type, an applied mathematician, if he talks on his current researches, is at a disadvantage *vis à vis* pure mathematicians and theoretical physicists, for they deal with ideas and with general theories while he is a solver of special problems. If he goes into the background of his work he finds he has trespassed into their fields. For that reason I have chosen to survey a wide field in a branch of applied mathematics which is less familiar than most others (and in which I am not in any sense an expert). The choice may be justified by the fact that, because one of the sessions of this congress is devoted to a discussion of the training of applied mathematicians, it might be of value to look at a fast growing subject to see if it can give us fresh material both for research topics and for teaching purposes.

2. The application of mathematics to the discussion of problems in biology is not new. In the past there have been applications to specific problems which have been successful in helping to elucidate the nature of the phenomena being discussed, but there is a great difference between occasional activities of this kind and the systematic development of mathematical biology. The need for such a development has been suggested several times, but especially in that remarkable and enjoyable book *Growth and Form* by D'Arcy Thompson **(1)**.

Only in one field of biology was there, until recently, any sizable mathematical theory—namely, in the study of the interaction of biological species in a population of organisms. This fundamental work was carried out mainly by Lotka **(2)** and Vito Volterra **(3)**.

The systematic study of the application of mathematics to a wide range of biological problems can properly be said to have begun by Nicholas Rashevsky whose early researches are admirably summarized in his book **(4)** on mathematical biophysics, the publication of which in 1938 focused the attention of the mathematical world upon the wealth of problems which the biological sciences could offer the applied mathematician. That same year saw the founding of the *Bulletin of Mathematical Biophysics*, a quarterly published by the University of Chicago Press under the editorship of Professor Rashevsky who happily continues in that office. This journal (from which purely statistical studies are excluded), which is the principal outlet for research in mathematical biology in the wider sense as well as theoretical biophysics, has probably been the greatest single influence on the development of the subject. The only rival claimant would be Professor Rashevsky himself, the steady stream of whose books and papers have made a significant impact and have ranged in scope from biology proper to a mathematical biology of social phenomena.

There is a fundamental difference between the approach adopted in the works of Lotka and Volterra on the one hand and that of Rashevsky's book on mathematical biophysics on the other. Lotka and Volterra, starting from direct observation, postulate certain general relations between organisms, and develop from them a mathematical theory of various biological processes involving such relations. They are dealing with the organic world as a whole, not with the detailed structure of each individual organism or of the relations existing between the vital parts of an organism and the world. On the other hand it is with these latter conditions that Rashevsky is most concerned in *Mathematical Biophysics*. It is to be hoped that, ultimately, the postulates of Lotka and Volterra can be derived by a fundamental biophysical study of the individual organisms.

These two branches of mathematical biology developing side by side may be compared with the parallel development of thermodynamics and of atomic theory.

3. Mathematics has played a fundamental role in the development of the physical sciences not because it provides a mathematical description of physical phenomena but because it gives us the possibility of setting up realistic models of the actual physical situation whose behaviour can then be treated by standard mathematical techniques. By the help of this analysis we attempt to gain a deeper understanding of the complex physical process to which the model refers and through that understanding to test various hypotheses by a direct appeal to experiment. The conclusions to be drawn from these experiments often enable us to build models of greater complexity whose analysis leads, in turn, to still further experiments. This interplay between mathematical theory and practical experimentation is a complex feedback process with which we are familiar through the study of physics. The very act of constructing a mathematical model may itself be fruitful in that it suggests the measurement of certain parameters occurring in the mathematical equations which describe the behaviour of the model. Even a crude model of a real situation has the merit that it is a clearly defined basis from which to discuss the design of new experiments and from which to seek new hypotheses. We may indeed make as our point of departure a highly idealized system which has no counterpart in nature, and we may go on not only studying the behaviour of those non-existent systems in some detail but also applying our conclusions to the real world. The whole success of classical mechanics and its applications to astronomy and engineering is a tribute to the usefulness of such unreal concepts as "point particle" and "rigid body."

Mathematics has not played such an important role in the development of biology and medicine. Biologists have tended, in general, to exhibit a certain amount of fear and mistrust towards mathematics and this is only in small part due to their lack of knowledge of the techniques of mathematics. It springs rather from the fact that too often in the past models of biological processes were either too realistic or oversimplified. If they were too realistic, their analysis defied the techniques (or the patience) of the mathematician, and if they were oversimplified they provided very little insight into actual biological phenomena.

The final goal of workers in this field is the establishment of a subject called "mathematical biology" which would bear the same relation to experimental biology as mathematical physics bears to experimental physics. In mathematical biophysics this might be taken to involve the derivations of all the properties of various kinds of living cell from a few basic axioms. Such an aim seems at the moment to be quite unrealistically high and, at least for the present, it is necessary to set up different sets of axioms for different branches of biology. One day it may prove to be possible to embrace all of these sets of axioms in a single common set. Here again the resemblance to mathematical physics is immediately obvious; heat, light, and electricity were developed as distinct disciplines until the present century.

It should perhaps be emphasized here that mathematical biology does not claim to "explain" life in terms of physical concepts. Indeed it is for the future to decide whether or not such a thing is possible. If it is, then it may well be for the reason that biology will have developed from basic axioms which are not necessarily reducible to concepts of physics as we know it today.

It is to be hoped that the study of mathematical biology will lead to the discussion of purely theoretical stituations with no immediately obvious "practical" significance. The importance of such developments is manifest in the history of physics. We might hope that there would be a parallel development in mathematical biology. No theoretical science can long survive without a firm experimental foundation but, on the other hand, no corpus of experimental facts can have real significance, unless it is capable of interpretation against a theoretical background. It is in this way that mathematics penetrates into the experimental sciences, but there is always a danger that it will become a mere tool of the experimentalist unless there is carried on a vigorous study of purely theoretical problems with an intrinsic interest of their own, irrespective of whether or not they appear to be immediately applicable to experimental results.

In mathematical biology, as in other branches of applied mathematics the important art to acquire is that of constructing models which are adequate to the physical situation under consideration and yet are capable of being handled mathematically. A fully rigorous approach will often lead to a quite absurd situation. For instance, in the simple problem in chemotherapy which we shall consider later (§ 6 below) the complete formulation of the problem would involve us in writing down the differential equations which express the conservation of matter taking into account the flood flow rate for each capillary, filtration and diffusion across capillary walls, and diffusion in the extracellular space, and formulating the geometrical relations which exist between all the capillaries in the region and between the extracellular space and the cells and capillaries. When we note that there are wide variations in the linear dimensions of individual capillaries and that in 1 sq. mm. of tissue there may be functioning as many as 190 capillaries we see that to attempt to formulate the problem on a fine-structure basis would result only in establishing the absurdity of such an approach. A macroscopic model indicating the broad outlines of the problem may not only be easier to handle mathematically but in the interpretation of experimental data may actually prove more useful.

The devising of successful models usually comes only as a result of the collaboration of a biologist and a mathematician, but for such a collaboration to be fruitful it is essential that the biologist should know something of the language and methods of mathematics. Above all he must have a sufficient knowledge of the methods of applied mathematics to appreciate that when an applied mathematician makes a statement about the physical world he is in reality making a statement about the model he has set up to represent the part of it relevant to the problem under discussion. He should learn, for instance, that in some problems the earth is treated as a massive point particle and in others as a semi-infinite solid with an infinite plane boundary!

This places upon university mathematics departments the responsibility of providing courses specifically designed for biology students, courses in which alongside the theory there are applications, and the situations being idealized are drawn from the biological sciences as well as from mechanics. Such a course might well be of value to many undergraduates to whom we have to teach calculus and who will *not* continue with the study of mathematics. Surely they can get an insight into the methods of applied mathematics as well by analysing a simple problem in biology as by doing simple problems in mechanics. Indeed it might do our aspiring specialist mathematicians a world of good to learn that applied mathematics embraces more than theoretical mechanics. The need is for elementary textbooks which will illustrate the various topics in mathematics with biological applications rather than the more conventional applications to geometry and dynamics. Recently an attempt has been made to provide a calculus text with this kind of approach **(5)** and other books of a like kind are being prepared by other authors.

Of course if the collaboration between biologists and mathematicians is to be successful at research level, applied mathematicians will have to learn something of the spirit and methods of biology. (Fortunately there are several texts available to facilitate this, one of the best being Canadian in origin **(6)**.) Then he will have to make a study of the field in which his particular problem lies. This sounds an exacting task but it is no worse than the one he would have to face if he were working in mathematical physics or econometrics.

4. In the last twenty-five years the development of mathematical biology has progressed rapidly. A number of highly sophisticated theories have been established but, of more significance to the experimental biologist, a great many observed facts, from fields ranging from cytology to experimental aesthetics and sociology, have been represented adequately in a quantitative way and in many instances new results have actually been predicted by the theory.

Because the best known examples of the application of mathematics to biological problems are in the field of experimental physiology, the casual observer may be tempted to think that the bulk of these are merely routine applications of calculus and differential equations. The theory of ordinary and partial differential equations does indeed play an important role in the mathematical analysis of certain biological problems but so do set theory, topology, and abstract algebra. Indeed, any pure mathematician who shares with Hardy the hope that none of the mathematics he does will be of any *use* to anyone is liable to have his aspirations frustrated by mathematical biologists. (Hardy's own name is well known to geneticists through a law which bears his name!)

Even when the tool being applied is one already well used the biological problems sometimes exhibit novel features not previously found in the study of mathematical physics or of econometrics. For instance, in a problem we shall consider later on, the distribution of a drug in the blood-stream, the built-in periodicity and time lag caused by the circulation of the blood leads to differential-difference equations involving partial derivatives with boundary conditions which themselves involve the time lag.

It might be of value here to list some of the biological problems which have been tackled by mathematical methods.

The earliest contributions were made in the discussion of problems of growth and form in individual animals and in populations and this work has continued **(7)**. Here the main tool is the theory of differential equations, but recently Ulam and others have considered the use of combinatorial methods in the solution of problems relating to patterns of growth.

A more recent development has been the systematic study of hydrodynamical problems of circulation and the transport of hormones, enzymes, and so on **(8)**, in which the main tools are the standard processes of mathematical physics, such as the theory of the solution of differential equations. The methods of the theory of differential equations also play a central role in the discussion of radioactive processes occurring within the body and of problems in physical biochemistry. A discussion of certain simple problems involving radioactive tracers is given in **(5)** and a fuller analysis of the difficulties encountered in the use of such tracers is given in **(9)**. There have been theoretical investigations on the circulatory system beginning with Otto Frank's classic work in 1899. Recent investigations which should be noted are Cope's elastic reservoir theory of the human systemic arterial system using data on aortic elasticity, Defares' theory of lung functions, and Bernard's mathematical studies on the interaction of respiratory gases with whole blood. (See the papers by Bernard listed in **(11)**.) Some of the simpler problems are discussed in **(5)** but an interested reader must consult the review article by Macey **(10)**. In the field of physical biochemistry there have been significant advances in the development of mathematical descriptions and analyses of metabolizing and endocrinological systems **(11)**.

In recent years there has been a great deal of activity in the development of information theory, mathematical logic, and combinatorial analysis and in their applications to problems in genetics, neurophysiology, and the synthesis of computer elements to simulate the brain. The classical theory of genetics has been described by

Fisher and Malecot **(12)**. Of more recent interest is the work of Rosen and others on the DNA–protein coding problem **(13)**.

Mathematical analysis of the operation of the nervous system including the brain has reached a high level of sophistication as is evidenced by Parts II and III of Rashevsky's book **(4)** and by Householder and Landahl's book **(14)**. Of interest to mathematicians there have appeared, since then, Landahl's work on the matrix theory of neural nets, the work of Rapaport and Shimbel on the probabilistic approach to this theory, and the work of McCulloch, Pitts, and Roberts on the Boolean algebra of neural nets. Some other recent work worthy of note is listed in **(15)**.

Recent advances in the design of calculating machines and of devices for the automatic control of industrial plant have led to a fresh discussion of the extent to which it might prove possible to construct mechanical or electrical models possessing properties which resemble the behaviour of living organisms. It is quite clear that in the study of such models we are far from the situation confronting the biologist—that of deducing the structure and function of living organisms by controlled observation of their behaviour—but it may be that the construction of these simpler physical models will help us to an understanding of the more complex biological situation. A whole new science of automata studies is growing up around the basic ideas first put forward by John von Neumann **(16)**. In these studies many new and interesting questions arise, concerned with the synthesis of computer elements to simulate the brain and with reliability problems (in which we ask if and how it is possible to design reliable machines, or organisms, using unreliable components). Most fascinating of all are the problems connected with self-reproduction: how can machines be designed that are capable of reproducing themselves? There is now an appreciable literature on these topics, the briefest extract of which is given in **(16)**.

The application of probability theory and statistical methods to the discussion of biological and medical problems has a long history. Statistical methods are too numerous (and sufficiently well known) to need mention here; the growth of the application of the theory of stochastic processes is perhaps not as well known but its development has been rapid. For instance, chapter IV of **(17)** is devoted entirely to biological applications of Markov processes, and the bibliography for that chapter (pp. 230–4) lists 98 items. In this field mathematical epidemiology has long been a flourishing subject, which now has an excellent introductory treatise of its own (**18**).

In discussing applications of probability it is interesting to note that, in a paper delivered to the Second Conference in the U.S.S.R. on the application of mathematical methods in biology, the distinguished statistician B. V. Gnedenko described the work of his school on the problem of making the diagnosis of heart diseases more accurate by the development of new instruments, by a statistical approach to illness and by the analysis of the logic by which diagnosis is carried out. Gnedenko's paper was followed by one by M. A. Kulikov in which he outlined a plan for the statistical study of the history of illnesses, and the study of electrocardiograms by means of harmonic analysis and showed that the ultimate object, the construction of an electronic machine having an extensive memory store containing previous observations on patients, was possible.

So far we have mentioned only applications of mathematics to specific biological problems and said nothing about the development of the subject which we earlier referred to as "mathematical biology." This is, of course, a much more difficult subject and one which we might expect would develop only after the solution of a great many more special problems. However, there have been significant developments in the establishment of such a general theory, important among which are those due to Rashevsky and his followers on "topological biology" (the application of topological arguments in the discussion of biological problems) and that due to Kerner on a form of statistical mechanics of interacting biological systems (**19**).

The use of mathematical methods in medical diagnosis is not as widespread as in the experimental biological sciences. Among applications of some interest we mention the use of mathematical techniques in the elucidation of electro- and ballisto-cardiography (**5**, p. 345, and **20**), in following the time course of an anaesthetic during anaesthesia (**5**, p. 353), in interpreting the results of measurements of cardiac output by the dye dilution method (**5**, p. 429), in using the Leyden heart function test (**5**, p. 540) and in clinical applications of cupolometry (**5**, p. 570). In surgery an important application is the reconstruction of spatial objects from the evidence provided by X-ray diffraction patterns. More recently there has arisen the fascinating study of how far medical diagnosis can be carried out by an electronic computer, which is, of course, fed clinical information in the appropriate form.

In the field of chemotherapy there are many interesting problems in which the concentration time-course of a drug in the body is studied. We shall look at two problems in this field: the first one (§ 5) is naive in its approach and uses only simple ideas of calculus, but is useful as an example of the kind of illustration we can use in an elementary course on calculus for medical students; the second (§ 6) is much more sophisticated and has the advantage that it brings us to the frontiers of knowledge in a fascinating field.

5. We shall consider first the calculation of the optimal dosage of drugs based on the simple assumption of van Gemert and Duyff (**5**, p. 255) who state that "in a general way at any given time the amount eliminated, destroyed or inactivated, and thus the decrease of concentration in the body fluids, is the greater as the amount present, and thus the concentration is greater. In the great majority of cases the velocity of elimination at any moment will be very nearly proportional to the concentration; in other words, the law of exponential elimination will hold, at least approximately." With this assumption we find that, if the concentration is initially (that is, at $t = 0$) c, then at a time t later the concentration is

$$c = c_0 e^{-t/\tau}$$

where τ is a constant which is called the *elimination time*. If we take this elimination time to be our unit of time, then

$$c = c_0 e^{-t}. \tag{1}$$

We now suppose that a dose c_0 is given at regular intervals of time t_0. It is then easy to show that the concentration of the drug in the body at the beginning of the nth interval will be

$$c_{n-1} = c_0(1 + e^{-t_0} + e^{-2t_0} + \ldots + e^{-(n-1)t_0}),$$

and at the end of that interval it will be

$$r_n = c_{n-1}e^{-t_0} = c_0(e^{-t_0} + e^{-2t_0} + e^{-3t_0} + \ldots + e^{-nt_0}).$$

Hence we have that as $n \to \infty$, $r_{n-1} \to r$ where

$$r = \frac{c_0}{e^{t_0} - 1}. \tag{2}$$

The maximum concentration attained immediately after a further dose c_0 will be

$$c_{\max} = r + c_0 = \frac{c_0}{1 - e^{-t_0}} \tag{3}$$

and from this moment onwards the concentration will decay according to the law

$$c = c_{\max} e^{-t} = \frac{c_0 e^{-t}}{1 - e^{-t_0}} \tag{4}$$

so that after a further interval t_0 it will have the value r, showing that after an interval t_0 we are back to the residual concentration given by equation (2). From this it follows that, once the concentration r has been established, the concentration will oscillate between $c_{\max}$ (at the instant when the dose is given), and r (the concentration just prior to the administration of the next dose). This means that the first (large) dose could be c_0+r followed by doses c_0 at regular intervals. The practical consequences of these results are formulated as follows by van Gemert and Duyff: "In those cases where it is essential that the concentration never falls below a given minimum level, while on the other hand, the allowable margin between this level and a level which may not be surpassed (for instance because of untoward side effects) is narrow, as in those where it is desired that a nearly constant concentration be maintained, it is advisable, once the residual concentration has practically reached its end value r, to give small doses at frequent intervals. In order, however, that this condition, where a given concentration is always surpassed, be present from the beginning of treatment (which is often the case, because 'subthreshold' concentrations may, in many cases, not only be ineffective, but harmful as with several chemotherapeutic agents and antibiotics) the first dose should equal the sum of the residual dose and the maintenance dose."

This is a very simple calculation and the result may not be in accord with observations. It should be noted that if there is a failure between theory and experience, it does not arise because we have adopted an approximate procedure in the course of the calculations but because the model we have adopted (as reflected in equation (1)) is too simple to account for the actual phenomenon occurring in the body fluid. As an example of a model of the same process but with more detailed structure we shall, in the next section, consider some recent work by Bellman, Jacquez, and Kalaba **(22)** on the mathematical theory of chemotherapy.

6. The first paper by Bellman *et al.* is described as being the first of a series in which the authors study qualitative and quantitative aspects of chemotherapy. The basic problem they aim at solving is that of determining the distribution of a drug in the organs of the human body after its injection into the blood-stream, and hence of determining chemical injections with optimal conditions.

They first consider a relatively simple system consisting of the heart and one organ, and proceed to set up a simple physical model adequate to describe the biological situation and capable of being analysed mathematically. The simplest model they use is shown in Figure 1 which refers to the situation when the drug diffuses into the extracellular space but does *not* enter the intracellular space. For a discussion of its validity on physiological grounds, the

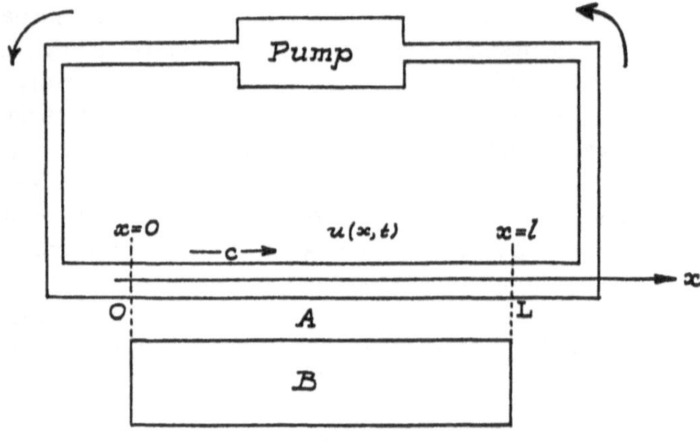

FIGURE 1.

interested reader is referred to the original paper. In this model the heart is represented by a pump, all the capillaries of the organ are conceptually lumped together into one capillary OL, of length l, in which the concentration of the drug (in mass per unit volume) at position x in the capillary at time t is denoted by $u(x, t)$. It is assumed that this capillary contains a volume R_p of plasma and that the volume rate of plasma flow in the capillary bed is c (assumed constant). In Figure 1, B denotes the extracellular space whose volume is denoted by R_e and in which it is assumed that the concentration of the drug is uniform and is denoted by $v(t)$. The region A denotes the region where exchanges between the blood and the cells of the region take place; it is assumed to be such that the surface area of contact between OL and B is A_e. If we take into account the motion of the blood in the circulation and diffusion of the drug into the extracellular space we readily see that the conservation equation can be written in the form

$$\frac{R_p}{l}\frac{\partial u}{\partial t} + c\frac{\partial u}{\partial x} = \frac{kA_e}{l}(v - u) \tag{1}$$

where k denotes the permeability constant for capillary walls (in units of lt^{-1}). On the other hand, if we examine the rate of change of concentration of the drug in the region B we arrive at the equation

$$R_e\frac{dv}{dt} + kA_e v(t) = \frac{kA_e}{l}\int_0^l u(x, t)dx, \tag{2}$$

valid for $t > 0$, $0 < x < l$.

To approximate to the conditions produced by an intravascular injection we assume that a square concentration pulse of magnitude u_0 and duration t_1 enters the capillary at $t = 0$. We also assume that a particle travels round the circulation from L to O in a time τ ($> t_1$) so that the boundary condition is of the form

$$u(0, t) = \begin{cases} u_0, & 0 < t \leqslant t_1 \\ 0, & t_1 < t \leqslant \tau \\ u(l, t - \tau), & t > \tau \end{cases} \tag{3}$$

and the initial conditions are

$$u(x, 0) = v(0) = 0. \tag{4}$$

The condition $\tau > t_1$ means that we are restricting our attention to injections of duration less than one circulation time. The model assumed will probably not describe accurately events of very short duration, particularly if the injection is of such short duration that the volume of plasma occupied by the drug (ct_1) represented a small fraction of the volume (R_p) of the capillary bed. For this reason we take $ct_1 > R_p$. Taking the two conditions together we find that we are restricted to considering cases for which

$$R_p/c < t_1 < \tau. \tag{5}$$

The problem of solving equations (1), (2) subject to the conditions (3) and (4) is of great interest and, because of the fact that the boundary condition (3) involves a time difference, of a different type from those normally encountered in mathematical physics. Assuming that the solution exists and is unique, it is a relatively simple matter to solve this problem formally by taking Laplace transforms with respect to the time variable t. The trouble arises at the end when we try to find analytically the inverse of the transform $U(x, s)$ of the concentration. For that reason Bellman *et al.* consider a simplified model in which the x-dependence disappears from the equations.

In the simplified model it is assumed that there is a certain concentration, $u_1(t)$, at the arterial end O of the capillary and another $u_2(t)$ at the venous end L, and that the diffusion into the extracellular space depends on the average $\frac{1}{2}(u_1 + u_2)$ and on $v(t)$. With these assumptions equation (1) is replaced by

$$\tfrac{1}{2}R_p\left(\frac{du_1}{dt} + \frac{du_2}{dt}\right) = c(u_1 - u_2) - kA_e[\tfrac{1}{2}(u_1 + u_2) - v], \tag{6}$$

equation (2) by

$$R_e \frac{dv}{dt} = kA_e[\tfrac{1}{2}(u_1 + u_2) - v], \tag{7}$$

and equations (3) and (4) by

$$u_1(t) = \begin{cases} u_0, & 0 < t < t_1, \\ 0, & t_1 < t < \tau, \\ u_2(t - \tau), & t > \tau, \end{cases} \tag{8}$$

and

$$u_1(0) = u_2(0) = v_2(0) = 0, \tag{9}$$

respectively. It can easily be shown that for this approximation to be valid we must have $kA_e/c < 2$, for otherwise $\dot{u}_2(0)$ would be zero and $u_2(t)$ would be negative for small values of t.

The system of equations (6) to (9) has been solved formally by Bellman *et al.* again using the method of the Laplace transform but again there is some difficulty in evaluating the contour integral by which the result is expressed. However, this system of ordinary differential equations is admirably suited for solution on a high speed digital computer. The function $u_1(t)$ is defined for $0 \leqslant t \leqslant \tau$ and $u_2(t)$ and $v(t)$ can be found over the same interval using the differential equations (6) and (7) and the initial conditions (9). Equation (8) enables us to find $u_1(t)$ for $\tau \leqslant t \leqslant 2\tau$ and so on. The presence of the time lag demands an increase in the size of the computer memory necessary to carry out the calculations, but it is otherwise quite simple. The model we have considered here is a very simple one; for a more complicated one greater demands would be made on memory space which might prove to be a limiting factor in the setting up of models sufficiently realistic to meet the biological situation.

An additional factor which is easily incorporated into this simple model is the possibility of the drug entering the cells themselves and there reacting chemically with substances within the cells to form new compounds. For instance, if the drug D reacts with the enzyme E to form a compound ED and if it is assumed that neither the enzyme E nor the compound ED diffuse out of the intracellular space but that there is an active transport process which tends to concentrate the drug intracellularly then equation (6) remains unaltered but equation (7) is replaced by a simple system of three simultaneous ordinary differential equations involving u_1, u_2, v, w (the uniform intracellular concentration of the drug), and z (the concentration of ED in the intracellular space), and the initial conditions are suitably modified.

This system of equations can also be handled best by the use of a high speed digital computer. Once the solutions of equations of this type have been studied exhaustively for wide ranges of the relevant parameter, it should be possible to construct input-output models for the main organs of the body (the model for the kidneys has to be modified by the addition of a sink to include the effects

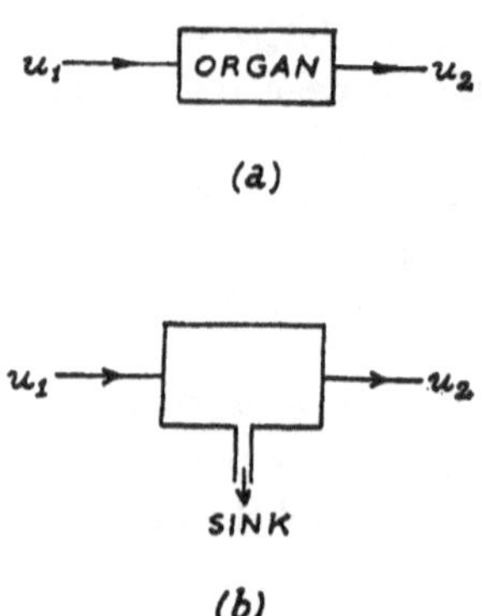

FIGURE 2. (*a*) Input-output model of a typical organ. (*b*) Input-output model of kidneys.

of urinary excretion) and from these units to construct a model of the whole body by forming a parallel arrangement in relation to the blood supply of the type shown in Figure 3.

It should be emphasized that in putting forward these models, Bellman *et al.* are not claiming that they can be treated by the methods of ordinary mathematical analysis; throughout they are thinking of the possibility of employing large high speed computers (either digital or analogue) to derive purely numerical results. They conclude their second paper by saying "We feel that the advent of the modern computing device will permit us to approach for the first time all types of biological problems which, because of their complexity and sheer magnitude, have previously been only sadly surveyed from a distance." The publication of their further work on the subject is eagerly awaited.

7. This has been an incomplete and inadequate survey of a fast growing subject but perhaps enough has been said to give a glimpse of the wealth of problems in biology available to an applied mathematician to widen his field of research interest and enliven and

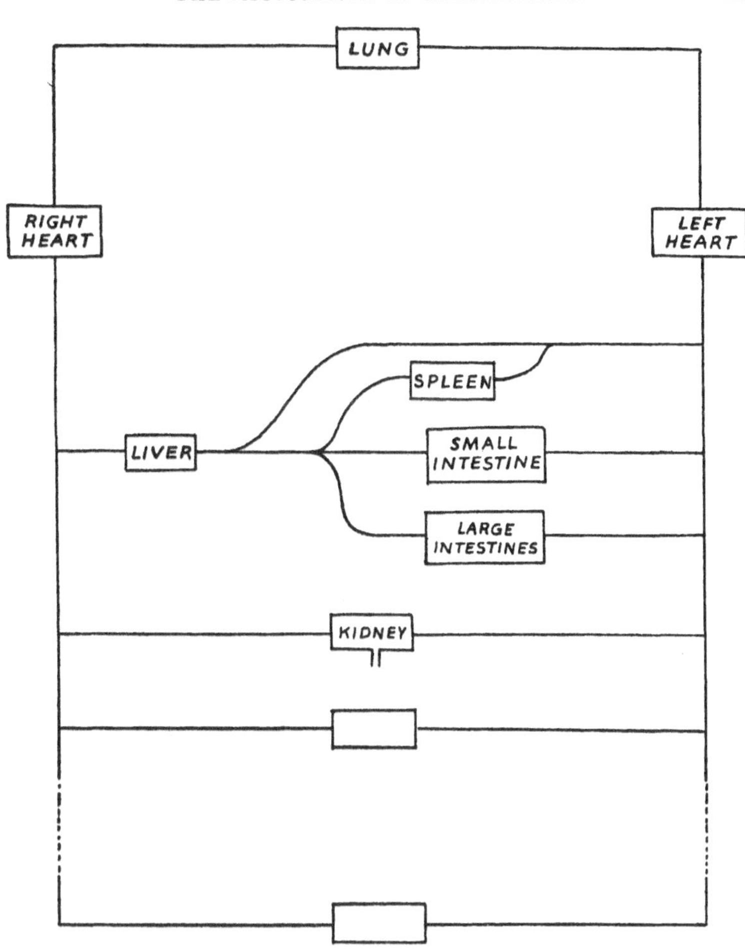

FIGURE 3.

enrich his teaching of the standard procedures of applied mathematicians. Almost all that has been said in this lecture is inspired by the philosophy and the writings of Nicholas Rashevsky, the founder of mathematical biology. Perhaps it is not inappropriate therefore for me to borrow words of his to express my own sense of inadequacy to deal with the topic on which I have elected to talk. At the end of the preface to the second edition of his great book Rashevsky writes: "An old oriental legend tells of a wizard who was asked a simple question by his wife and did not know the answer to it. His wife strongly reprimanded him for this, saying, 'You are the counsellor to our king who pays you much money for your knowledge; are you not ashamed of not knowing such simple things?' 'My dear,' replied the wizard, 'the king pays me only for what I know. If he were to pay me for what I don't know, all the treasures of his kingdom would not be sufficient for that purpose.' "

REFERENCES

1. D'Arcy W. Thompson, On growth and form (Cambridge, 1917).
2. A. J. Lotka, Elements of physical biology (Baltimore, 1925; reprinted as Elements of mathematical biology (New York, 1956)).
3. V. Volterra, Variazioni e fluttuazioni del numero d'individui in specie animali conviventi (Venezia, 1927).
   ——— Leçons sur la théorie mathématique de la lutte pour la vie (Paris, 1935).
4. N. Rashevsky, Mathematical biophysics: Physico-mathematical foundations of biology (Chicago, 1938; re-issued in two volumes by Dover Publications (New York, 1960)).
5. J. G. Defares and I. N. Sneddon, An introduction to the mathematics of medicine and biology (Amsterdam, 1961).
6. C. H. Best and N. B. Taylor, The living body, 4th ed. (New York, 1960).
7. S. J. Huxley, Problems of relative growth (London, 1932). W. E. le Gros Clark, and P. W. Medawar (eds.), Essays on growth and form presented to D'Arcy W. Thompson (Oxford, 1945). J. W. Cragg and N. W. Pirie, The numbers of man and animals (Edinburgh, 1955).
8. T. Rosenberg, On accumulation and active transport in biological systems, Acta Chem. Scand., *2* (1948), 14.
   ——— The concept and definition of active transport, Symp. Soc. Exp. Biol., *8* (1954), 27.
   H. H. Ussing, The relation between active ion transport and bioelectric phenomena, Monograph, Instituto do Biophysics, Universidade de Brasil, Rio (1955).
   E. J. Harris, Transport and accumulation in biological systems (London, 1956).
   C. S. Patlak, Contributions to the theory of active transport, Bull. Math. Biophysics, *18* (1956), 271; *19* (1957), 209; *23* (1961), 173.
   J. L. Stephenson, Theory of transport in linear biological systems, Bull. Math. Biophysics, *22* (1960), 1113.
   O. Jardetzky, On the definition of active transport, Bull. Math. Biophysics, *22* (1960), 103.
9. A. K. Solomon, The kinetics of biological processes; Special problems connected with the use of tracers, Adv. in Biol. & Med. Physics, *3* (1953), 65.
10. R. Macey, Studies on the relations between pressure and flow in the circulation, Bull. Math. Biophysics, *17* (1955), 169.

11. H. Branson, A mathematical description of metabolizing systems, Bull. Math. Biophysics, *8* (1946), 159; *9* (1947), 93.
J. Z. Hearon, A note on the integral equation description of metabolizing systems, Bull. Math. Biophysics, *15* (1953), 15.
L. Danziger and G. L. Elmergreen, Mathematical models of endocrine systems, Bull. Math. Biophysics, *19* (1957), 9.
S. Roston, Mathematical representation of some endocrinological systems, Bull. Math. Biophysics, *21* (1959), 271.
S. R. Bernard, Mathematical studies of the interaction of respiratory gases with whole blood, Bull. Math. Biophysics, *22* (1960), 391; *23* (1961), 1.
F. Heinmets and A. Herschman, Qualitative analysis of metabolic processes, Phys. Med. Biol., *4* (1960), 238; and Bull. Math. Biophysics, *23* (1961), 69.
S. Roston, Glucose metabolism—a "mixed-cycle" system, Bull. Math. Biophysics, *23* (1961), 99.
12. R. A. Fisher, The genetical theory of natural selection (Oxford, 1930).
G. Malecot, Les mathématiques de l'hérédité (Paris, 1948).
13. G. Gamow and M. Ycas, Statistical correlation of proteins and amino acid composition, Proc. Nat. Acad. Sci., *41* (1955), 1011.
G. Gamow, A. Rich, and M. Ycas, The problem of information transfer from the nuclei acids to proteins, Adv. in Biol. & Med. Physics, *4* (1956), 23.
R. Rosen, The DNA-Protein coding problem, Bull. Math. Biophysics, *21* (1959), 71.
——— Some further comments on the DNA-Protein coding problem, Bull. Math. Biophysics, *21* (1959), 289; *22* (1960), 171.
——— A quantum-theoretic approach to genetic problems, Bull. Math. Biophysics, *22* (1960), 227.
14. A. Householder and H. D. Landahl, Mathematical biophysics of the central nervous system (Bloomington, Ind., 1945).
15. J. Tobias, Biophysical aspects of conduction and transmission in the nervous system, Ann. Rev. Physiol., *21* (1959), 299.
A. R. Adolph, Feedback in physiological systems: an application of feedback analysis and stochastic models to neurophysiology, Bull. Math. Biophysics, *21* (1959), 195.
H. D. Landahl, A note on mathematical models for the interaction of neural elements, Bull. Math. Biophysics, *23* (1961), 91.
H. White, The formation of cell assemblies, Bull. Math Biophysics, *23* (1961), 43.
16. J. von Neumann, The general and logical theory of automata *in* Cerebral mechanisms in behaviour (New York, 1951).
——— Probalistic logics and the synthesis of reliable organisms from unreliable components *in* Automata Studies, Annals of Mathematics Studies No. *34*, ed. C. E. Shannon and J. McCarthy (Princeton, 1955).
R. Rosen, On a logical paradox implicit in the notion of a self-reproducing automaton, Bull. Math. Biophysics, *21* (1959), 387.
17. A. T. Bharucha-Reid, Elements of the theory of Markov processes and their applications (New York, 1960).
18. N. T. J. Bailey, The mathematical theory of epidemics (London, 1957).
19. N. Rashevsky, Topology and life: In search of general mathematical principles in biology and sociology, Bull. Math. Biophysics, *16* (1954), 317.
——— Note on a combinatorial problem in topological biology, Bull. Math. Biophysics, *17* (1955), 45.

——— Some theorems in topology and a possible biological implication, Bull. Math. Bipohysics, *18* (1956), 31.

——— What type of empirically verifiable predictions can topological biology make? Bull. Math. Biophysics, *18* (1956), 173.

——— A note on the geometrization of biology, Bull. Math. Biophysics, *19* (1957), 201.

——— Remark on an interesting problem in topological biology, Bull. Math. Biophysics, *19* (1957), 205.

——— A comparison of set-theoretical and graph-theoretical approaches in topological biology, Bull. Math. Biophysics, *20* (1958), 267.

——— A note on the biotopology of reproduction, Bull. Math. Biophysics, *20* (1958), 275.

——— A note on topological biology, Bull. Math. Biophysics, *21* (1959), 97.

——— A set-theoretical approach to biology, Bull. Math. Biophysics, *21* (1959), 101.

——— Contributions to relational biology, Bull. Math. Biophysics, *22* (1960), 73.

R. Rosen, A relational theory of biological systems, Bull. Math. Biophysics, *20* (1958), 245; *21* (1959), 109.

——— The representation of biological systems from the standpoint of the theory of categories, Bull. Math. Biophysics, *20* (1958), 317.

E. Trucco, A note on Rashevsky's theorem about point-bases in topological biology, Bull. Math. Biophysics, *18* (1956), 65.

——— Topological biology: a note on Rashevsky's transformation *T*, Bull. Math. Biophysics, *19* (1957), 19.

E. Kerner, A statistical mechanics of interacting biological species, Bull. Math. Biophysics, *19* (1957), 121.

——— Further considerations on the statistical mechanics of biological associations, Bull. Math. Biophysics, *21* (1959), 217.

——— On the Volterra–Lotha principle, Bull. Math. Biophysics, *23* (1961), 141.

20. R. H. Kesselman and A. Grishman, The physical rationale of ballisto-cardiography, Bull. Math. Biophysics, *19* (1957), 247.

G. C. K. Yeh and J. Martinek, Comparison of surface potentials due to several singularity representations of the human heart, Bull. Math. Biophysics, *19* (1957), 293.

J. Martinek, G. C. K. Yeh, and H. de Beaumont, The inconsistencies of present-day mathematical–physical methods pertaining to current generators in volume conductors used in electrocardiography, Bull. Math. Biophysics, *21* (1959), 61.

S. Roston, Mathematical formulation of cardiovascular dynamics by use of the Laplace transform, Bull. Math. Biophysics, *21* (1959), 1.

D. A. Brody, J. C. Bradshaw, and J. W. Evans, The elements of an electrocardiographic lead tensor theory, Bull. Math. Biophysics, *23* (1961), 31.

21. S. G. Vandenberg, Medical diagnosis by computer: recent attempts and outlook for the future, Behavioral Sci., *5* (1960), 170.

22. R. Bellman, J. A. Jacquez, and R. Kalaba, Some mathematical aspects of chemotherapy, Bull. Math. Biophysics, *22* (1960), 181, 309.

www.ingramcontent.com/pod-product-compliance
Lightning Source LLC
Chambersburg PA
CBHW050257110726
47898CB00007B/2441

9781487573300